TELECOMMUNICATIONS

TELECOMMUNICATIONS

Present Status and Future Trends

by

Robert F. Linfield

National Telecommunications and Information Administration
U.S. Department of Commerce

Advanced Computing
and
Telecommunications Series

NOYES DATA CORPORATION
Park Ridge, New Jersey, U.S.A.

Copyright © 1995 by Noyes Data Corporation
Library of Congress Catalog Card Number: 94-31267
ISBN:0-8155-1368-2
Printed in the United States

Published in the United States of America by
Noyes Data Corporation
Mill Road, Park Ridge, New Jersey 07656

Transferred to Digital Printing, 2010

Printed and bound in the United Kingdom

Library of Congress Cataloging-in-Publication Data

Linfield, R.F.
        Telecommunications : present status and future trends / by Robert
F. Linfield.
            p.   cm. -- (Advanced computing and telecommunications series)
        Reprint. Originally published: Washington, D.C. : U.S. Dept. of
Commerce, 1993.
        ISBN 0-8155-1368-2
        1. Telecommunication. I. Title. II. Series.
TK5101.L57   1994
004.6--dc20   94-31267
                                                            CIP

Preface

The purpose here is to define the present and examine the future of telecommunications over the next ten years. Emerging and anticipated products and services are viewed from both a technical and a social impact perspective. Systems including those providing voice, data, images, video, and integrated services are investigated in terms of technical feasibility, standardization, and global applications. Networks and concepts discussed include: LANs, MANs, WANs, wireless networks, switched multimegabit data service (SMDS), ISDN, B–ISDN, synchronous transfer mode (ATM), and synchronous optical networks (SONETs). The information gleaned from this study is summarized in a series of tables and charts that characterize the critical parameters of various switching and transmission systems and concepts as well as the network architectures. The major architectural concepts and systems expected to have critical impact on the future telecommunications infrastructure are presented along with important issues expected to affect their evolution.

With the addition of Chapter 10 (Addendum—Telecommunication Trends Update), this book is brought up–to–date as of mid–1994.

To whatever degree I have managed to bring together this comprehensive overview of telecommunication trends, at least some of the credit goes to Messrs. V.J. Pietrasiewicz, W.J. Pomper, and J.A. Hull who provided technical suggestions. In addition, helpful discussions were held with many other members of the staff at the Institute for Telecommunication Sciences.

Robert F. Linfield

Notice

Certain commercial equipment, instruments, services, protocols, and materials are identified in this report to adequately specify the engineering issues. In no case does such identification imply recommendation or endorsement by the National Telecommunications and Information Administration, nor does it imply that the material, equipment, or service identified is necessarily the best available for the purpose.

The materials in this book were prepared as accounts of work sponsored by the U.S. Department of Commerce. On this basis the Publisher assumes no responsibility nor liability for errors or any consequences arising from the use of the information contained herein.

Contents and Subject Index

1. Introduction

"One thing about the past, it's likely to last!"

Ogden Nash

The telecommunications industry in the United States is growing approximately 12% per year. Its contribution to the Gross National Product (GNP) by the year 2000 is expected to be greater than 20%. Today this $335 billion dollar industry is approximately 6% of the GNP. Computers and computing services are approximately 10% of the GNP. The U.S. represents approximately one third of the world market in telecommunications.

This report emphasizes telecommunications and what one might expect this important technical area will look like ten years from now. Recognize, however, that Ogden Nash's quotation above will continue to apply. The telecommunications infrastructure even today still uses many analog mechanical switches, plain old telephone service (POTS), teletypewriters, and

1.2 kb/s data modems. In looking at future trends, as is done here, it should always be kept in mind that the imbedded plant has lots of inertia and replacing the old with the new is a long process. The report begins by examining the past and the current posture of telecommunications.

1.1 Yesterday and Today

Twenty-five years ago there were no fiber optics, no microelectronics, no video recorders, no compact discs, and no cellular telephones. The public switched telephone network was metallic with analog switches and copper transmission facilities. Systems and services like FM stereo radio, color TV, cellular radio, T-carrier, audio cassettes, cable TV, and communication satellites were just beginning to emerge. The telephone industry, dominated by AT&T and Western Electric, was extensively regulated. There were no public packet switched networks, no stored program controlled switching, no personal computers (PCs). The fields of computer based operating systems, distributed processing, and cellular telephones were still to come.

Today, nearly everyone has a PC in the workplace and two FM radios at home. Fiber optical transmission, digital switching, and communications satellites are taken for granted. Local area networks (LANs) interconnect a multitude of personal computers. Each desktop PC has as much raw processing power as a fairly large main-frame computer had just ten years ago. Fiber optical transmission facilities proliferate. Network evolution continues to migrate toward information packetization, packet switching, and digital transmission with more and more features and functions under customer control. Today's distributed computing networks handle bursty, high-speed, high-volume traffic with low delay.

In the United States, deregulation and procompetitive initiatives have superseded many of the traditional monopolistic practices. An increased reliance on a market economy has led to greater efficiencies in networking and innovative products are providing higher standards of living.

Voice and data communication growth has expanded globally with some 700 million telephone terminals and data terminals in existence worldwide (Brule and Ebert, 1990). Figure 1-1 indicates the phenomenal growth in worldwide voice and data terminals over the past 15 years.

A similar increase in local area networks (LANs) to interconnect these terminals, metropolitan area networks (MANs) to interconnect LANs, and wide area networks (WANs) are

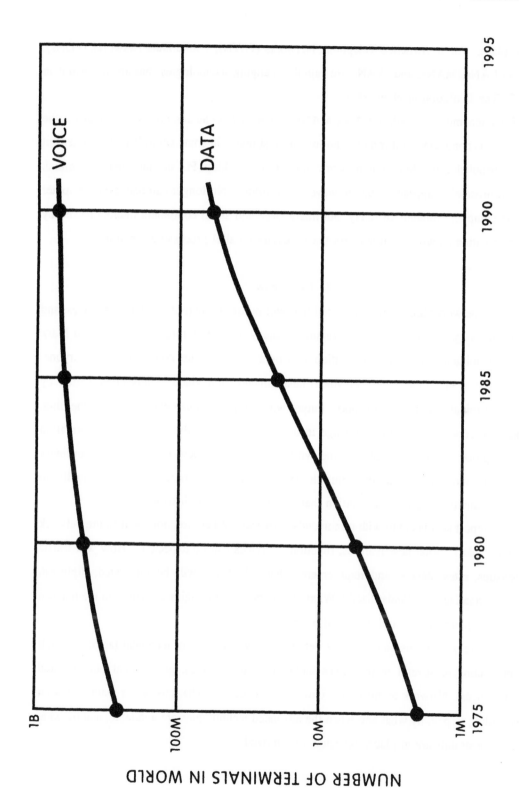

Figure 1-1. Growth of voice and data communications (Elgen, 1990).

common place. Public Switched Telephone Networks (PSTN), Public Data Networks (PDNs), as well as LANs, MANs, and WANs are rapidly changing technologies that are discussed in Section 7 (The Evolution of Networks).

The phenomenal growth of LANs, MANs, and WANs is causing profound changes in the way business is conducted. Information no longer must reside in some centralized computer but can be distributed throughout the network and in the desktop PC or multimedia terminal. Distributed processing appears to be the wave of the future. No longer can one large computer handle the complex problems of today, but many large and small computers interconnected by telecommunication networks can perform the necessary parallel processing functions.

1.2 Tomorrow

Future networks are being envisioned that will boost productivity, enhance services, and enrich our lives by extending social interaction to many diverse cultures. As Mayo and Marx (1989) have pointed out, these information networks of the future could provide everyone, anywhere, and at any time access to voice, data, images, and video in any combination by plugging a terminal into a universal port. Future wireless systems can even eliminate the plug. The aim is to educate, entertain, exchange ideas, and manipulate data and create information around the globe. It is apparent that the world is entering a new era where extraordinary advances in technology are leading us into the information age where there will be virtually no limit to the amount or type of information that can be accessed or distributed.

This report is concerned with tomorrow's networks. What directions will future network architectures take? What new features, functions, services can be expected? How will traffic characteristics, rates, delays, and requirements change? What will be integrated, digitized, packetized, privatized, or standardized? What are the new technologies currently on the horizon that may be implemented? And what will be their impact?

Future service provisioning will be determined by a number of technical factors, as well as the interaction of several major forces including market forces, government policies, and global influences. Network architectures will be influenced by what customers want and what they are willing to pay, by what and when the advanced technologies are available, and by what regulatory constraints are in place, expected, or removed.

A major force moving the U.S. into this information age has been the merging of the telecommunications and computer industries under a generally favorable deregulatory environment. Computers are the information processing systems while telecommunications provides means for information dissemination. Computers are also the brains of the networks, providing control and management functions.

This computer and telecommunications merger raises a number of jurisdiction issues, both national and international. Standards is one such jurisdictional issue. National interests may sometimes conflict with global interests. World markets must be taken into account when considering what standards should be adopted in the future.

Other factors that will influence the network architectures of the future are indicated in Figure 1-2. In addition to national and international standards, factors such as technology, government policy, and users' needs are discussed in detail in subsequent sections of this report.

The evolutionary directions that future network architectures may take or are taking are depicted in Figure 1-3. In each case, the path to the future begins with an existing structure. This is an important point to remember throughout this report. Seldom does a new creative concept displace or replace the existing structure instantaneously. Rather, each concept evolves over time as economics and demand allows. Embedded facilities are gradually replaced with new technologies (e.g., fiber replacing copper for long-haul transmission, and digital replacing analog for switching and transmission) over periods of time and depending on cost benefits and amortization schedules. These changes result from government regulations, new technologies, and the market place, with technology being the major agent of change. Whatever induces the change, the result is that the network infrastructure is seldom static but always a mix of the old with the new. This is demonstrated in Figure 1-4, showing the level of deployment of major long-haul switching technologies over the years. (The asynchronous transfer mode (ATM) is a packetized multiplexing scheme discussed in Section 3.4 of this report.) This constantly changing structure may be expected to continue in almost every aspect of the network. Other examples are given throughout this report.

The fact that networks of the future must continue to coexist with present networks for some time adds complexity to the standards-making process, to interoperability, and to the design of the telecommunication switching, transmission, and control facilities.

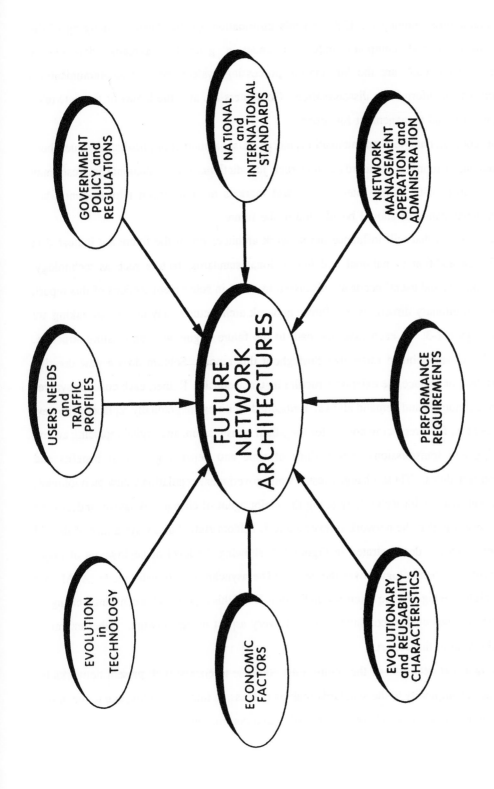

Figure 1-2. Factors influencing architectural definition.

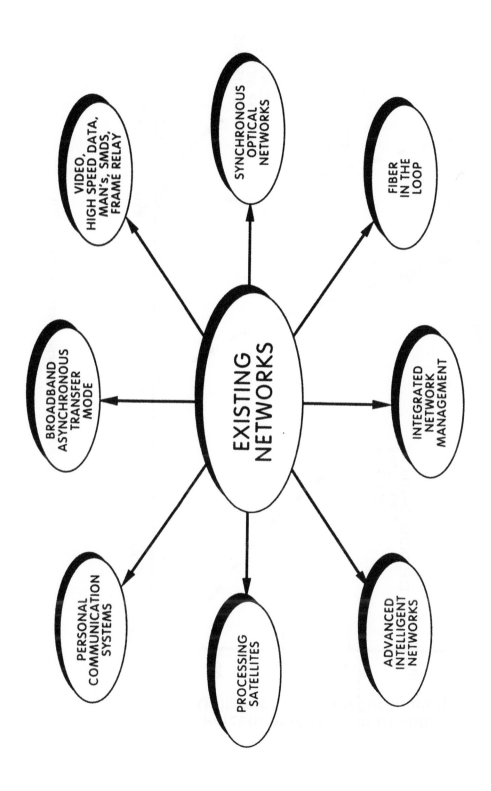

Figure 1-3. Evolutionary directions in telecommunications.

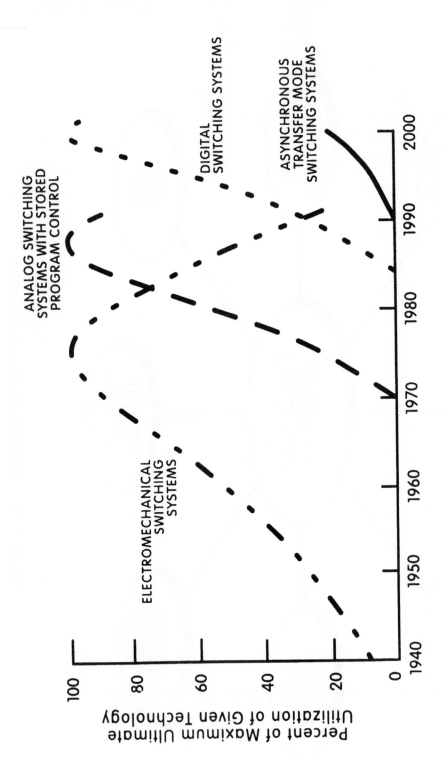

Figure 1-4. Switch technology deployment (from Katz, 1990).

1.3 Report Synopsis

The topics of Sections 2 through 9 of this report are interrelated in a complex way, as shown in Figure 1-5. This is because numerous factors and forces influence the evolving telecommunications infrastructure. For example, information is a key economic resource and our society is becoming increasingly involved in information services. Information has one distinctive character defined by its "half life," the time for a given pattern of information to lose 50% of its value or meaning. Thus, the use of telecommunications to disseminate information fast and accurately lies at the heart of social activity. This coming information age is discussed in Section 2. As the United States moves into a knowledge-based, service-oriented economy, the telecommunications infrastructure becomes an important factor for economic development. Some important new technologies impacting this infrastructure are described in Section 3. The directions it takes in the future depend on several forcing functions. These include government regulations and standards described in Section 4, along with market forces and users needs as covered in Section 5. All of these factors—technology, government policy, market forces, and users needs—combine to generate new products, systems, and services as described in Section 6. These in turn have a major affect on the actual network infrastructures as described in Section 7. The information contained in Sections 2 through 7 serves as the basis for the summary Section 8 concerning major trends and issues. Finally, Section 9 closes with conclusions and recommendations. A reference list is provided in Section 10.

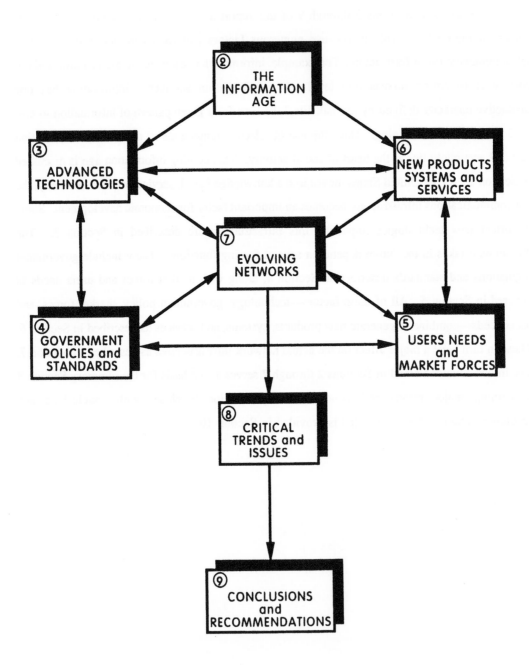

Figure 1-5. Interrelatedness of topics covered in this report.

2. The Coming of the Information Age

> *"The opportunities presented by universal information services are stunning. While the telephone extended the reach of the human voice, universal information services promise in the decades ahead to extend the reach and capability of the human mind."*

C. L. Brown, Chairman of AT&T, 1985

In order to assess the future course of telecommunications in the U.S. and to develop future architectural concepts, it is useful to examine the critical events that have taken place in the past. Table 2-1 indicates a division of the evolution of telecommunications into four epochs-- the age of creation (1850-1900), the age of ubiquity (1900-1950), the age of diversity (1950-2000), and the age of information exploration (2000-2050). The age of creation is when the critical underlying inventions occurred (e.g., the telegraph, the telephone, and the wireless). Near the end of this period, the Bell patents expired and competition in telecommunications really began.

In the early 1900's, a universal service was the major goal. Over 90% of American households had radios by 1950 and telephones by 1970. The Communications Act of 1934 that created the FCC was a driving force. Until mid-century, the FCC regulated communication service on the assumption that it could best serve the public as a monopoly. Universal service would be provided by rate averaging and affordable service would result from the economies of scale.

Around 1960, the government began to consciously follow a different policy--that of promoting competition in the industry. Milestones in this era include the Carterphone decision in 1968 (expanding the terminal equipment market), MCI decision in 1969 (resulting in a specialized carrier industry), the Department of Justice's antitrust suit in 1974 (leading to divestiture of the Bell Operating Companies (BOCs) in 1984), and Computer Inquiries I, II, and III (to define jurisdictional responsibilities as the field of communications and computers converged).

The complete evolutionary picture must include an age of information exploration covering the period 2000-2050. This period applies the tremendous information base that is available to everyone in order to expand knowledge. Universal service and information access will be the foundation of this new information age.

11

Table 2-1. Periods of Telecommunications Developments

EVOLUTIONARY AGE	YEAR	TECHNOLOGY	POLICY
AGE OF CREATION	1850	1842 - Telegraph 1876 - Telephone 1886 - Wireless Telegraph	1893 - Bell patents expire (Start of competition)
AGE OF UBIQUITY	1900	1921 - Mobile Phone	1907 - AT&T refuses interconnection 1910 - Mann-Elkins Act (ICC Regulation) 1913 - Kingsbury Commitment (Interconnection Required) 1927 - Radio Act 1934 - Communications Act 1949 - AT&T refuses interconnection 1949 - AT&T antitrust suit
AGE OF DIVERSITY	1950	1950 - Microwaves 1954 - Transistor 1955 - CATV 1957 - Satellites 1964 - Carterphone 1970 - Fiber Optics 1971 - Microelectronics 1988 - ISDN 1994 - PCS Cell Relay SONET 1994 - Intelligent Network 1	1956 - Consent Decree 1962 - All Channel TV Receiver Act 1966 - Computer Inquiry I 1968 - Equipment Interconnection 1969 - MCI Microwave Approved 1970 - Satellite Policy 1974 - AT&T Antitrust Suit 1984 - BOC Divestiture 1988 - ONA Plans
AGE OF INFORMATION EXPLORATION	2000	• Advanced Intelligent Networks (AIN) • B-ISDN via ATM & SONET • Integrated workstations (voice, images, data, video) • Information centers • Distributed processing • Wireless access • HDTV • Photonic Switching • Networks of Networks	• Further Deregulation • ONA • Global Standardization • OSI
	2050	?	?

Developments in the field of information technology have already paved the way for a new age when the collection, generation, and dissemination of information is paramount. New applications of information services pervade all areas of society from offices and manufacturing plants to schools and homes. And these applications are constantly increasing. Computers are becoming more and more a pervasive influence in our daily life. The origination, storage, manipulation, and distribution of information is critical to the success of any modern business venture. Distribution has caused a need for expanding network capacity resulting in a tremendous bandwidth explosion.

This so called "information age" is actually upon us now, but its full potential has yet to be realized because people are in the midst of changing the very foundations upon which this new age is based--namely telecommunications. The power to create and manipulate information is critical, but information exchange at a distance, i.e., telecommunications, provides the means to make it accessible to all. New services that were unthinkable just a few years ago, like voice messaging, cordless telephones, mobile systems, imagery, high definition TV, and more, are now available to users. New features and functions continue to be added constantly. Changes in the telecommunications infrastructure today are so dramatic that it is nearly impossible to provide a true course for the future. One can only look at where the world is now and where its wants to go.

People have just embarked on the information age and have only caught a glimpse of its ultimate potential. According to NTIA (1991) "The power to create and manipulate information is critical to capturing the promise of the information age, so also is the ability to move that information from point to point. This latter capability is, of course, provided by telecommunications, and it is why the U.S. telecommunication infrastructure is commonly referred to as the "highway" of the information age." This "highway" is expected to play a critical role in improving domestic economy in the U.S., the welfare of its citizens, and their ability to be competitive in the future marketplace of the world.

More will be said about the major features and functions of the network infrastructure for future decades in the following Sections 3 through 7. These trends and issues are then summarized in Section 8.

3. The Promise of Advanced Technology

"-- carrying human voice over copper wires is impossible and even if it is possible the thing would have no practical use"

Newspaper Editorial in 1870

The world has come a long way since commercial telephone service began shortly after the above quote appeared. The advances in telephony and related industries continue to emerge. Table 3-1, which is reproduced from Bonatti et al. (1989), depicts the evolution of telecommunications since the 1950's. Notably absent are newer technologies like personal communication systems and the asynchronous transfer mode.

Eight years ago Mayo (1985) described what he called three killer technologies: integrated circuits that killed the vacuum tube, fiber optic transmission that is replacing copper wires and cable, in some instances, and software that provides a new functionality and control. Today, all three continue to impact the industry, but new "killers" have been introduced. These include intelligent networks, controlled by artificial intelligence and expert systems, that are replacing many network management functions formally performed by humans, and ultimately could affect the entire infrastructure of telecommunication networks; photonics, that could push transmission and multiplexing speeds to the terabit per second region by the year 2000, and at the same time be used in switching arrays ending the need for electronic conversion; wireless technologies, that may free us all from the proverbial umbilical cord by providing universal services on a mobile-global basis; broadband switching, multiplexing, and transmission systems, that will replace today's narrowband world with new information age services like high definition television and multimedia desktop terminals; and finally, high-speed global interconnectivity using both hardware and software technologies that can provide significant increases in processing power, and networking capability to the user.

The following subsections describe these and other technologies that are expected to impact the telecommunications architecture. Section 7 then covers the specific networks that will use these advanced technologies and possibly others still on the drawing boards.

Table 3-1. Fifty Years of Telecommunications Network Evolution (from Bonatti et al., 1989)

	1950s	1960s	1970s	1980s	1990s
Switching	• X-BAR	• Stored Program Control (SPC)	• Digital Toll	• Local Digital Switching	• Wideband Switching
Transmission	• Analog Radio • Coax	• T-Carrier	• Digital Cross-Connect Systems • Digital Loop Pair Gain Systems • Digital Radio	• Fiber • Digital Circuit Multiplication Equipment	• Loop Fiber
Signaling	• In-Band SF and MF Signaling		• Out-of-Band Common Channel Toll Signaling	• Signaling-Based Service Utilizing Network Database	• Global End-to-End Signaling (CCS 7)
Customer Premises Equipment	• Private Branch Exchange (PBX) • Modems	• Automatic Call Distribution	• SPC PBX • Intelligent Terminals	• Digital PBX • LAN's • PC's • Facsimile	• Wideband Switch • Workstations
Data Communication Technology	• Modems		• Packet Switching • Circuit Switching	• Bandwidth Management Systems	• Asynchronous Transfer Mode • Wideband Packet Switch
Network Capabilities	• Automatic Alternate Routing		• Common Channel Signaling	• Dynamic Nonhierarchical Routing	• Worldwide Flexibility
Network Operations Systems	• Automatic Message Accounting	• Traffic Network Management	• Mechanized-Testing, Alarm Surveillance Traffic Admin	• Customer Control • Expert Systems	• Integrated Network Management • Automated International OPS
Public Switched Service	• Direct Distance Dialing (DDD)	• International DDD	• Free Calling Service ("800" Service)	• Long Distance and International Competition • Customer-Controlled "800" Service • ISDN	• Broadband ISDN
Private Network Service	• Tandem Tie Trunk Network • Private Lines	• Private Networks With Shared Switching	• Electronic Tandem Networks Switched Voiceband ($\leq$ 2.4 kbits/s)	• Software Defined Networks (SDN)	• International ISDN
Data Communication Service	• Voiceband Data Service	• Switched Voiceband Data ($\leq$ 1.2 kbits/s) • Nonswitched Voiceband ($\leq$ 9.6 kbits/s)	• Nonswitched Digital Data Services (56, 64 kbits/s) • Private Packet Switching • Public Packet Switching	• Switched Digital Service (56 kbit/s) • Nonswitched Data (2, 45 Mbits/s) • OSI	• Wideband Data-Switched (up to 150 Mbits/s) -Nonswitched (1.2 Gbits/s) • International Directory

3.1 Artificial Intelligence (AI) and Expert Systems

The potential for implementing AI technology into telecommunications, network management, operations, administration, and maintenance is almost unlimited but virtually untapped. According to Amarel (1991):

> "As a result of a series of advances in artificial intelligence, computer science, and microelectronics, we stand at the threshold of a new generation of computer technology having unprecedented capabilities. The United States stands to profit greatly both in national security and economic strength by its determination and ability to exploit this new technology."

AI technology offers advantages in software (e.g., languages such as LISP and programming advances) and at the same time reduces the costs of the hardware needed to support the software. AI machines could, for example, permit a single user to control large amounts of dedicated memory and processing resources to enhance overall processing power. The principal application of this new technology is expected to be network management and controls.

AI technology could also be used to plan and manage future telecommunications facilities, provide automatic advisory programs, and lead to a new generation of powerful computer-based design and manufacturing systems.

3.2 Photonics

Future computing and switching systems may use optical technologies for storage, processing, and switching as well as the transmission of information. This is a major research challenge today. The following subsections describe some of the applications of photonic technology beginning with the most current application--the transmission of signals over optical fiber.

3.2.1 Optical Fiber Transmission Systems

Until recently, the bulk of long-range communications was point-to-point microwave with some help from satellites. The development of inexpensive efficient optical fiber has revolutionized the transmission industry. Fiber is an almost ideal medium for transporting high-bit-rate information because of its enormous capacity. At the same time, it occupies no radio spectrum, freeing the spectrum for other users.

The advantages of optical fiber are so great that it may even be used in some mobile radio systems and personal communications networks to provide the primary links between the fixed stations. Radio systems would only provide the short-range flexible link to mobile users. Since, the radio link is so short, very low-power portable transceivers become feasible for user convenience (Matheson, 1992, unpublished).

Optical fiber transmission systems have an enormous capacity for carrying information at low cost. Recent comparisons indicated typical capacities of 10^3 b/s for copper wire, 10^6 b/s for coaxial cable, and 10^9 b/s for fiber. The fiber capacity continues to increase each year due to technology improvements. Although fiber still only accounts for a small percentage of the mileage of copper, it has more than doubled copper networks' capacity.

In one decade, the AT&T optical fiber systems in service increased in capacity from almost 700 simultaneous calls to 48,000 calls over a single fiber pair. This corresponds to an increase in data rate from 45 Mb/s to 3 Gb/s (Ekas, 1991). According to Warr (1991), by the year 2000, Tb/s transmission speeds (10^{12} b/s) appear likely.

A typical coaxial cable weighs two hundred times as much as a fiber. In addition, fiber systems have low power requirements (10 mW is typical), are not disrupted by lightning, have no crosstalk problems, are relatively secure from interception, and are insensitive to many environmental effects.

Although fiber has been replacing copper wires and cable in many parts of the network, on the subscribers' side of the local telephone exchanges, (i.e., the local loops) copper still dominates. The replacement cost for the "last mile" (e.g., to the home or office building) is still excessively high because there is so much copper to be replaced. The current thrust for dealing with existing local loop cabling (plant) is to increase the copper's capacity by using compression techniques. The increase in capacity achieved by compression is limited and may not be sufficient for future video services.

Regarding the use of fiber in new introductions of cable, more and more fiber-to-the-home (FTTH), and fiber-to-the-curb (FTTC), is being installed. According to Shumate (1989), the cost of installing broadband FTTH in 1989 was between $5K and $10K; which was three to four times the cost of providing telephone and TV cable combined ($1.5K to $2.5K). These costs were expected to come down as demand for broadband services increase and the volume of fiber and components produced increases. The average installed costs of fiber and copper access lines as a function of time was given by Fahey (1991), who estimated that by 1995/1996 fiber will be competitive in case-by-case installations. The decreasing cost of fiber and increasing cost of copper is illustrated in Figure 3-1. This estimate may be conservative, however, if information presented at the October 1992 Newport Conference on Fiber Optics Markets is correct. It was indicated that in 1992 the circuit-mile cost of fiber was equal with that of copper. As more fiber cable is installed, the cost will decrease even further, and the cost of copper will increase due to the lower volume of copper cable used as seen in Figure 3-1. Five years ago this same conference reported that industry was reluctantly installing long-haul fiber optic cable. No one was sure that the broadband capability was necessary. Today this need is driving the installation of fiber cable. It was predicted that there will be at least ten more years of significant fiber cable installation. The 1997 volume of fiber cable is projected to be seven times that of 1992. Presently, undersea fiber cable for international service (to the former Eastern Block countries) is the most active fiber growth area. The next most active market is fiber-in-the-loop, also called fiber-to-the-home.

Fiber transmission systems provide the basic structure needed for B-ISDN which is intended to offer users a flexible, two-way medium and provide future broadband services such as switched video (see Section 7.3).

A fiber-based network known as the Fiber Distributed Data Interface (FDDI) has been standardized by ANSI. Applications for FDDI include its use (1) as a high-speed (100 Mb/s) backbone to interconnect low-speed LANs, (2) for connecting processors to other processors and I/O devices, and (3) for efficient access to wide area networks (WANs). The FDDI concept is discussed further in Section 7.2.

Installing fiber may not always be economically feasible, so compression techniques are used to reduce capacity requirements of conventional transmission media. The following examples demonstrate what can be achieved.

A T1 carrier operated at 1.544 Mb/s normally transports 24 two-way 64 kb/s voice circuits. Using 32 kb/s Adaptive Pulse Code Modulation (ADPCM), this same T1 carrier can transport 48 voice circuits. Today, a pair of optical fibers operating at 500 Mb/s is equivalent to over 300 T1 carriers and can therefore carry 14,000 voice circuits. In the near future, the potential number of voice circuits increases to approximately 5.6 million, if 16 kb/s digital voice

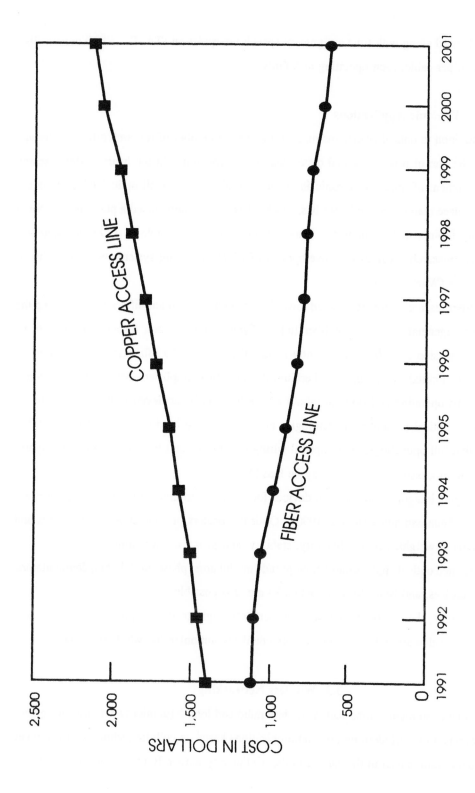

Figure 3-1. Average installed first cost of fiber and copper subscriber loops (Fahey, 1991).

processing is used along with time assignment speech interpolation (TASI), and 10 optical fiber pairs are used per cable, each operating at 5 Gb/s.

3.2.2 Other Photonic Applications

In addition to optical fiber transmission, there are a number of research efforts underway to increase the rate of processing and disseminating information. At the heart of this research is the development of optical-electronic integrated circuits (ICs), or all-optical ICs, that could ultimately replace many semiconductor ICs. AT&T has already announced a photonic switching system using lithium niobate devices. This system could be available in 1995 (Anonymous, 1991). Other research involves the development of ATM switching systems with 100 Gb/s to 1 Tb/s capacity (Warr, 1991).

Another research effort is directed toward developing optical amplifiers for increasing the capacity and regenerator spacing in fiber links. Typical transmission distance using optical-electronic conversion amplifiers is currently about 50 km for a 565 Mb/s circuit (equivalent to 120,000 voice circuits). Using cascaded erbium-doped optical amplifiers, Millar (1991) predicts 2 Gb/s systems operating at 1,000 km (about 621 miles) can be achieved in the near future.

Digital-optical computing technologies are described by Medwinter and Taylor (1991). They note that "all opticallogic" with optical wiring is very difficult to achieve but that a hybrid optoelectronic processor architecture may be feasible.

Emerging integrated optical (IO) devices for use in communication satellite repeaters are discussed by Anunasso and Bennion (1990). Optical repeaters provide substantial savings and improvements in weight, volume, linearity, stability, and power consumption.

Integrated optical applications for amplitude modulators, phase modulators, demodulators, switching matrices, and beam forming networks are also possible.

Ultimately, the use of the physical properties of light to do the processing that is now done in the time-domain with microprocessors could revolutionize the whole network.

3.3 Wireless Networking

Wireless communications such as mobile radio and long-haul microwave links have been around for some time. Modern mobile communications based on cellular technology to alleviate spectrum congestion began in the early 1980's. Cellular systems rely on a network of cell sites and transceivers to provide the mobile service. According to Rappaport (1991), there were over 6.3 million cellular-telephone users as of September 1991. Due to price and size reductions, these mobile systems have grown approximately 50% per year, but are reaching saturation due to customer demand and channel capacity limitations. Two different digital technologies are under development to increase cellular systems' capacities, namely: time division multiple access (TDMA) and code division multiple access (CDMA). Both have certain advantages and disadvantages. However, CDMA appears to be able to expand system capacity by at least a factor of 10 over the analog cellular systems (Miska et al., 1992; Madrid et al., 1991; and Viterbi, 1991).

Cordless telephones (CT) used in the home are just wireless extensions of up to 1,000 feet to the existing telephone network. A year ago there were 30 million CTs in the United States.

Cordless networking for personal computers is under development. Miska et al. (1992) and Madrid et al. (1991) describe wireless business systems including cordless phones and cordless networking for personal computers. Radio packet networks are already in use for transmission of 2-way data messages. Personal communication systems (PCSs) for voice and data are on the horizon. The Federal Communications System has recently (1992) allocated frequencies in the 1.8 GHz to 2.2 GHz band for PCS, wireless PABXs and LANs. Figure 3-2 illustrates the cellular market trend in the U.S. as forecasted by Madrid et al. (1991). Wireless LANs and some PCS concepts are discussed in Section 7. The PCS is considered as a major emerging system for future applications in Section 8.

3.4 Broadband Switching and Transmission

In the field of telecommunications, the term "broadband" characterizes the capability of various products, systems, and services. One definition would characterize any network that is capable of supporting transmission bit rates above 1.5 Mb/s as broadband.

Broadband switching and transmission systems are based on what is commonly called "fast packet" technology. Fast packet switching is a term that has been applied to a number of switching processes ranging from 1 or 2 Mb/s to 100's of Mb/s. Fast packet switching by this definition includes both frame relay and cell relay. Fast packet switching advantages include

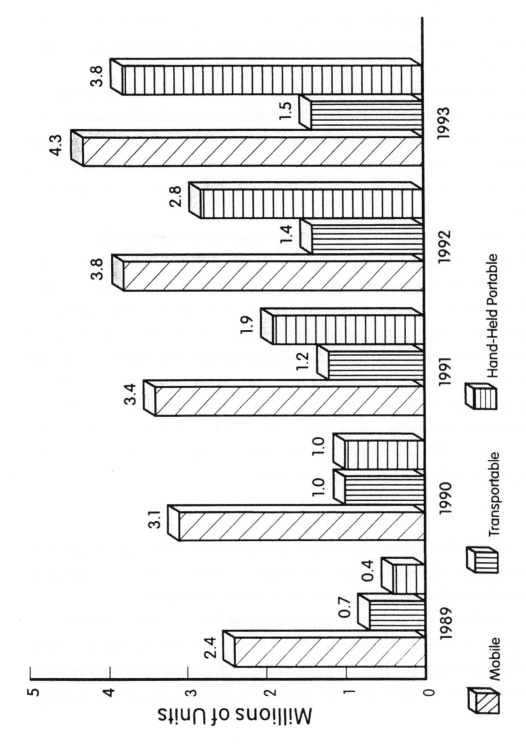

Figure 3-2. Cellular market trends in the United States (Madrid et al., 1991) .

- flexible and efficient user access,

- variable bandwidth information rates,

- true integration of all information services for transmission and switching,

- rapid support for new services.

To achieve fast packet switching at video rates, switches need low delay (1 ms per node), high interface rates (e.g., 150 Mb/s) and large capacities (more than 10 million packets per second). By comparison, many packet switches used today in public data networks have node delays greater than 20 ms, interface rates of about 48 kbit/s and capacities less than 10,000 packets per second. This delay and capacity is unsuitable for most voice or video services.

Fast packet switches will be implemented with dedicated hardware using very large scale integrated (VLSI) circuits. One possible switch design, called the Banyon network is shown in Figure 3-3. It consists of 12 interconnected 2 x 2 switch elements in 3 stages. Each element can either switch straight through or exchange inputs. The elements are controlled by routing bits in the packet header which are set equal to the destination output port number. There is a single path between any input and any output which is determined by the output port number. Each bit of the routing field in the packet header indicates how the appropriate element of each stage should switch. A "one" indicates a switch of the input line to the upper output line, while a "zero" indicates a switch of the input line to the lower output line. The path through each element is held until the complete packet has been passed. The path selection is thus a simple step-by-step procedure based solely on the packet header. The switch can be implemented entirely in hardware and has many channels that can be processed in parallel. VLSI technology makes this concept feasible.

The following subsections describe some of the important switching, multiplexing, and transmission systems that are currently implemented or planned for the near future.

3.4.1 Frame Relay

Frame relay is a type of fast packet technology for transporting user data traffic over digital transmission links. Recommendation I.233 of the CCITT defines frame relay as an ISDN bearer service, which is based on extensions to Recommendation Q.922 called the "core aspects."

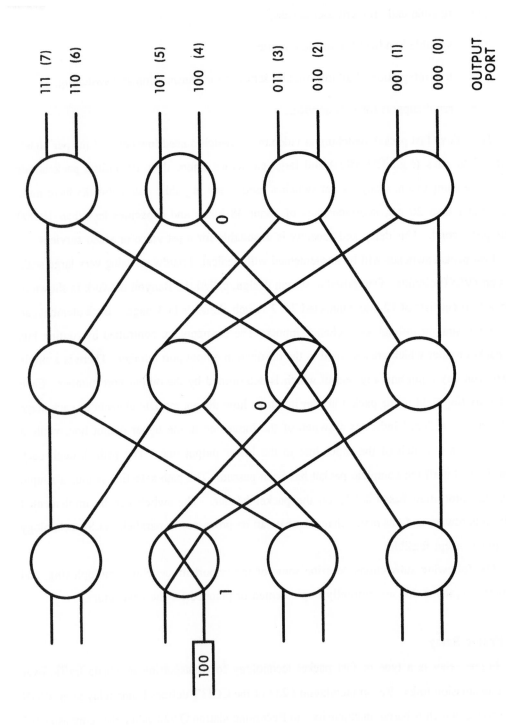

Figure 3-3. Fast packet switch, binary routing network.

It is expected to eventually replace X.25 for comparable data communication applications. The X.25 protocol is designed to provide error-free delivery of user data over high error-rate links while frame relay relies on digital links that are essentially error free. The X.25 definition covers layers 1, 2, and 3 of the OSI model, while frame relay is defined for layers 1 and 2. Table 3-2 is a comparison of frame relay and cell relay technologies, which collectively define the concept of fast packet. Cell relay technology, which is used in the B-ISDN environment, is explained in Section 3.4.2. Korpi (1991 a and b) evaluates the frame relay concept in more detail.

Table 3-2. Frame Relay and Cell Relay Compared

Frame Relay	Cell Relay
A broadband technology	A broadband technology
Optimized for data	Optimized for multimedia (i.e., voice, data, video)
Based on variable-length message units (frames)	Based on fixed-length message units (cells)
Intended for speeds below 45 Mb/s	Intended for speeds above 45 Mb/s
Typically implemented in software	Typically implemented in hardware
Can interwork with cell relay systems	Can interwork with frame relay systems

Frame relay systems enclose user data into variable-size larger packets called frames, which typically may be 1,000 bytes or more in length. By contrast, in cell relay systems, user data is transported using fixed-size packets called cells. Every cell is identical in length although some may be full, some partly full or empty. Frame relay systems are more compatible with today's data networks operating at 1.5 or 2.0 Mb/s, whereas cell relay systems support very high speeds (45 to 600 Mb/s) and serve multimedia applications efficiently.

Frame relay is designed to interface with ISDN's primary rate interface at 1.5 Mb/s (North America) and 2 Mb/s (Europe) but could be enhanced to 45 Mb/s in the near future. Faster packet technologies using cell relay, such as Switched Multi-megabit Data Service (SMDS) and the Asychronous Transfer Mode (ATM), operate over optical fiber links and will be used for the higher speeds. These concepts are described in subsequent sections.

3.4.2 Cell Relay

Cell relay may ultimately replace frame relay. It is also referred to as cell switching (McQuillan, 1991). Cell relay uses fixed-size packets of data called cells. Each cell includes 48 bytes of user information and 5 bytes of overhead (4 for segmentation and reassembly and 1 for control). The Switched Multi-megabit Data Service (SMDS) and Asynchronous Transfer Mode (ATM) are also based on the cell relay concept, as is the IEEE 802.6 standard for public MANs which uses a distributed-queue-dual-bus (DQDB) architecture (Section 7.2). ATM may be used for WANs. Along with the synchronous optical network (SONET), ATM provides the standardized switching and transmission base for B-ISDN. Details of ATM and SONET are given in later sections. Their application in the B-ISDN infrastructure is covered in Section 7.3.

3.4.3 Switched Multi-megabit Data Service (SMDS)

The SMDS is a connectionless network-layer service (CLNS) provided by commercial carriers for data transport. It is a near-term, public, wide area network extension of existing and LAN and MAN services. It may be used to interconnect other high-speed networks (e.g., FDDI token ring LANs), at transport rates ranging from 1.5 to 100 Mb/s on a tariffed as-used basis. Presently, it is being implemented by local exchange carriers only for intra-LATA connectivity (mostly in major metropolitan areas). National service is anticipated in the 1996-1997 time frame.

The SMDS switch provides routing, sequencing, and billing functions. A SMDS router at the customer's interface provides OSI link and network layer protocol functions plus packet segmentation and reassembly. Because SMDS is a switched service, it can reduce the number of access lines required when compared to using private leased lines.

SMDS and frame relay are competitive servers in some areas. Frame relay is used, either in public or private networks, for virtual private line services at DS-1 rates, whereas SMDS is a public, dynamically-switched service between SMDS subscribers at DS-1 and DS-3 rates.

3.4.4 Asynchronous Transfer Mode (ATM)

The ATM is a standardized method of information transfer in which the information is multiplexed into fixed-length packets called "cells" and transmitted according to each user's instantaneous need. The short cell length of 53 octets (48 octets of information plus a 5-octet header) minimizes cell delay. A single 150 Mb/s fiber carries approximately 44 cells in a 125 μs frame, sufficient for high-quality video. At lower speeds (e.g., voice), users are multiplexed. An ATM switch differs from the conventional digital switch only at the functional level. The ATM switch uses a self-routing switching fabric as opposed to a software-controlled switching network. For a detailed description of some architectures for self-routing ATM cells see Daddis and Lorng (1989).

In summary, ATM

- provides a standard format for B-ISDN

- combines circuit and packet switching capabilities

- provides bandwidth-on-demand to meet any users applications

- supports multiuser architectures whereby many users can simultaneously share the network

- combines with SONET to provide transport of all B-ISDN services.

ATM uses two distinct connections called virtual path and virtual channel. The relationship between virtual channel, virtual path, and transmission path is shown in Figure 3-4 (a). A transmission path may carry several virtual paths and each virtual path may carry several virtual channels. Virtual paths and virtual channels may be switched according to the virtual path identifier (VPI) and the virtual channel identifier (VCI) located in the header of each cell (Figure 3-4 (b)). The virtual channel provides the logical connection between users while the virtual path defines the route the cell takes between source and destination. This dual-connection scheme reduces total overhead requirements and simplifies traffic flow control since this can be done at the virtual-path level. The 5-octet cell header format contains VPI, VCI, payload type, cell loss priority, error control, and flow control information. The 48-octet information field carriers the user's data. The B-ISDN standard specifies transmission rates of 155.52 and 622.08 Mb/s for ATM. All cells associated with an individual virtual channel and

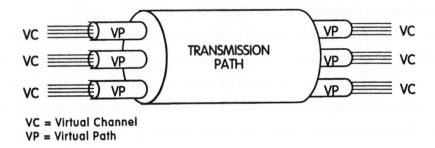

VC = Virtual Channel
VP = Virtual Path

a) Relationship between virtual channel, virtual path and transmission path.

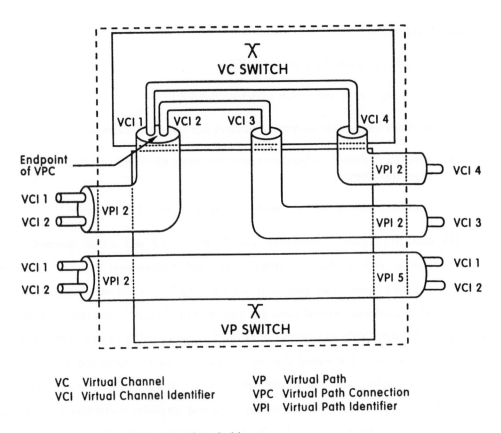

VC   Virtual Channel   VP   Virtual Path
VCI   Virtual Channel Identifier   VPC   Virtual Path Connection
                                   VPI   Virtual Path Identifier

b) Virtual channel/virtual path switching

Figure 3-4. The ATM transport network.

path connection are transported along the same route through the network. Cell sequence is preserved for all virtual channel connections. Figure 3-5 shows the OSI layered model for ATM. For a detailed description, see CCITT (1988b, d).

3.4.5 Synchronous Optical Network (SONET)

Some observers of the telecommunications evolution have made the analogy that SONET is to Broadband what T-1 Carrier was to Digital. SONET is the new family of optical transmission channels for speeds from 45 Mb/s to Gb/s. Just as T1 ports exist on customer premise equipment today, many telecommunications experts feel that SONET ports will be common features on equipment by 1997. The CCITT version of SONET is known as the Synchronous Digital Hierarchy (SDH). The relationship between SDH and SONET is shown in Table 3-3.

SONET defines standard interconnect line rates ranging from "STS-1" at 51.840 Mb/s to "STS-48" at 2.48832 Gb/s. When an STS-N signal is transmitted, the resulting optical signal is called an optical carrier OC-N. The number of 64 kb/s channels in each optical carrier is also indicated in Table 3-3.

The STS-1 frame format contains 810 octets and is transmitted over a 125 µs period. This results in the 51.84 Mb/s line rate for STS-1 as shown in Table 3-3. Multiples of STS-1 signals form STS-N signals by synchronously interleaving bytes from N STS-1 signals. Note that the lowest SDH level is STM-1 operating at 155.52 Mb/s. This corresponds to the SONET rate STS-3.

The STS-1 can be divided into subrates called virtual tributaries (e.g., VT 1.5 Mb/s) and used to carry the North American digital hierarchy up to and including level DS-3 as shown in Table 3-4. VTs come in four sizes: VT 1.5, VT2, VT3, and VT6.

A detailed description of the SONET concept is given by Ballart and Ching (1989). Its role leading to the development of B-ISDN is described by Stallings (1992) and the universal network node interface (NNI) for the SDH and SONET is described by Asatani et al. (1990). The SDH standard development was initiated by the CCITT Study Group XVIII in 1986 (CCITT, 1988a, c, and e). The IEEE (1992) special issue on gigabit networks provides a comprehensive set of papers concerning gigabit transmission systems and technologies including SONET and SDH.

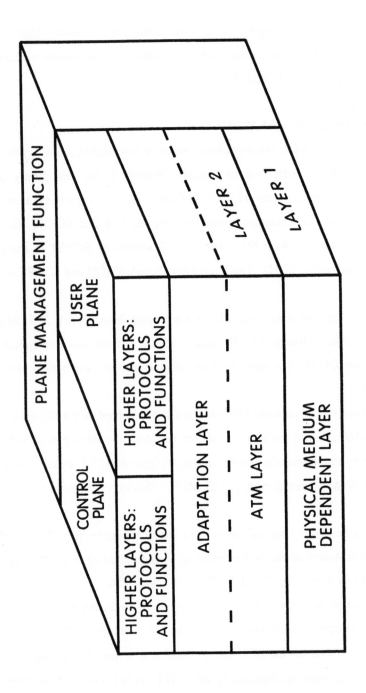

Figure 3-5. OSI layers for ATM.

Table 3-3. SONET Standard Digital Channels

SDH Designation	SONET Designation	Line Rate (Mb/s)	OC Level	64 kb/s Channels
------	STS-1	51.84	OC-1	672
STM-1	STS-3	155.52	OC-3	2,016
STM-4	STS-12	622.08	OC-12	8,064
STM-8	STS-24	1244.16	OC-24	16,128
STM-16	STS-48	2488.32	OC-48	32,256

Table 3-4. North American Digital Hierarchy

Digital Signal Levels	Transmission Rate	Number of T1 Equivalents	Digital Transmission Facilities
DS-4	274.176 Mb/s	168	T3M
DS-3	44.760 Mb/s	28	T3
DS-2	6.312 Mb/s	4	T2
----	3.152 Mb/s	2	T1C
DS-1	1.544 Mb/s	1	T1
DS-0	.064 Mb/s (64 kb/s)	1/24	---

As a high-capacity, fiber-based transmission system, SONET (or the SDH) may be expected to provide the future infrastructure for high-speed applications replacing the T-carrier technologies. The SDH will support B-ISDN using the asynchronous cell transmission of ATM.

3.5 Satellites

Satellite networks have advantages over terrestrial networks because (1) they are accessible from any place, (2) cost is independent of user to user transmission distance or intervening terrain, (3) they are ideal for broadcasting transmissions over large coverage areas and for communications to mobile users, (4) they can be time shared, (5) they can transmit and switch high-speed digital data (10's of Mb/s) to many points economically, (6) they can be implemented and reallocated quickly because no rights-of-way or cable laying are required, and (7) additional receiving sites do not add to distribution costs. Important applications include communications for undeveloped countries and the ocean areas.

Traditionally satellites have been viewed as radio repeaters with a very long, 22,300 mile, microwave hop. Satellite earth stations interface with terrestrial networks. Recently, advanced-technology satellites have been proposed that incorporate antennas with multiple spot beams and on-board switching systems to interconnect the spot beams to dynamically alter circuit capacities between locations. Such systems operating over different frequency bands could support the growth of satellite communication services and alleviate congestion problems (Wright et al., 1990).

Modern satellite systems meet business requirements for high-rate (~ 100 Mb/s) data transmission as well as video and voice communications. Applications include reservation systems for airlines or hotels, credit card verification, video teleconferencing, and inventory management systems. Transmission rates ranging from 9.6 kb/s to 2 Mb/s can be allocated on demand depending on the application.

Very small aperture terminals (VSATs) appeared in the 1980's, and today are used increasingly by business to provide direct connections of user data, voice, and video terminals. They serve as an integrated Wide Area Network (WAN), bypassing traditional terrestrial networks including the local loop and the long-haul facilities. There are several advantages to VSAT networks over leased terrestrial facilities. These include service reliability, single point of contact (VSAT provider), customer control, configuration management, and lower cost. A typical VSAT network today usually includes many remote user terminals and a single central hub. Transmission rates to the hub are on the order of 128 kb/s using time-division multiple access (TDMA). The hub transmits to all VSATs in a broadcast mode using time division multiplexing (TDM). Such a system typically can support 31 in-bound TDMA streams per out-bound TDM stream. This TDMA/TDM technology provides interactive data exchange with fast response, data file transfer, and even digitized voice.

Other time sharing techniques including single-channel-per-carrier (SCPC) are used with Ku-band VSATs. A VSAT network may include several thousand VSATs. By joining networks, over 100,000 VSATs can be served.

VSATs can also be used to communicate with unmanned sites for supervisory control and data acquisition. They also perform load management functions for the electric utility industry (Jaske, 1991).

Satellite-based personal communications systems (PCS) have also been proposed. One such concept called Iridium would use a constellation of 77 low earth orbit (LEO) satellites to provide worldwide coverage. Iridium uses cell-forming antennas and radio relays located on the satellite rather than on the ground. Otherwise, the system is similar to terrestrial cellular systems. An overview of Iridium is given by Grubb (1991). Other mobile satellite systems are being discussed as listed below.

- Local Space and Qualcomm are proposing the Globalstar system, which has 48 satellites in LEO (orbiting at about 800 miles above the earth). Subscribers would receive direct satellite services through hand-held terminals. The Globalstar system, using spectrum in the L-band or S-band, would provide coverage over most of the earth's land masses and major ocean areas, but not over the polar regions. Inter-satellite links are not used.

- TRW is proposing a system called Odyssey, consisting of 9 satellites at mid-earth orbit (MEO). These satellites would orbit approximately 6,000 miles above the earth. Plans call for the system to use the same satellite bands as the previous systems and may also include Ku-band.

- American Mobile Satellite Corporation (AMSC) and Telesat Mobile Inc. are proposing a geostationary system using Ku-band spectrum. The status of the proposed system is uncertain, pending action at the Federal Communications Commission (FCC).

- The European Satellite Agency is studying a proposed satellite system which uses highly elliptical orbits, from a perigee of about 100 miles to an apogee of about 700 miles. This system would provide coverage over Europe.

The World Administrative Radio Conference (WARC) meeting in Barcelona, Spain, allocated spectrum in the L-band (1.5 GHz) and S-band (2.5 GHz) for these satellite applications. Intersatellite links may operate in the Ku-band.

In 1987, there were over 200 satellites proposed or operating in the geostationary orbit according to Jansky and Jeruchim (1987). The investment in these systems is several billion dollars.

A satellite system in a synchronous orbit has several advantages as a very long (22,300 mile) radio repeater. Operating in the microwave band (4 and 6 or 12 and 14 GHz), the satellite can receive signals from properly oriented earth station antennas and relay them almost any place on earth in the coverage area. The transmission loss is therefore independent of the distance or intervening terrain between earth stations. Such satellites are ideal for broadcasting purposes since they can provide wide earth coverage, and propagation delay is not a factor in broadcast applications. Satellites can be time-shared on demand or prescheduled. They can carry tens of megabits per second and switch high-speed data to many points economically. Finally, satellite systems can be implemented quickly and reconfigured easily if desired.

Table 3-5, from Pelton (1988), compares the pertinent characteristics of satellites in synchronous orbit to an advanced fiber-optic cable system. Pelton also discussed the feasibility of using satellites in an ISDN environment and emphasized the potential cost savings and innovative services inherent in using both space and terrestrial communications.

Table 3-5. Satellite Versus Fiber-Optic Cable

	Advanced Satellite	Advanced Fiber-Optic Cable
System Availability	99.98%	99.98%
Bit Error Rate (BER)	10^{-7} to 10^{-11}	10^{-7} to 10^{-11}
Capacity (bits/sec)	1 gigabit to 3.2 gigabits	840 megabits to 2.5 gigabits
Transmission Delay	250 ms	Under 50 ms
Typical end-to-end transmission time	350 to 800 ms	200 to 700 ms

3.6 Hardware and Software Technologies

Much of this report is devoted to hardware technologies that interface with the traditional transmission media such as copper, optical fibers, and free space. Hardware components such as switches, multiplexers, transmission devices, and their associated computers are controlled and operated by software, however. Software technologies include languages of various types, operating systems, management systems, artificial intelligence, protocols, coding schemes, data base management systems, and the like. These software technologies are essential elements of any network. They make the hardware elements usable, permit the programming of control processors, manage the network's information base, and provide the interworking capabilities via protocols and expert systems. Protocols ensure the efficient transfer of information. Expert systems apply knowledge to manage and operate the network, monitor alarms, diagnose faults and automatically take corrective action to reconfigure topology, or switch in new components - all in essentially real time.

Software for controlling future networks such as B-ISDN must meet reliability, availability, and flexibility requirements far beyond those achievable today. Software modularity and reusability will provide a graceful evolution as new features and functions are added. See, for example, Vickers and Vilmansen (1987).

Digital networks today use hardware based on silicon technology for information processing, for performing switching functions, for storage elements, and for interfacing structures. Copper has been a major transmission medium but is being replaced by optical fiber. The capacity of a single-mode fiber has continued to increase year by year. By the year 2000 it could exceed 100 Gb/s as predicted by Bourne and Roth (1985). See also Section 3.2 on Photonics. Warr (1991) notes that semiconductor technology tends to follow certain laws and has for several decades. One example given is that memory and logic component density is expected to increase by a factor of 10 every five years through the end of this century. Magnetic storage density is increasing by a factor of four every five years and optical storage densities are increasing at a slightly slower rate. Finally, Warr notes that the cost/performance or size/performance of functions such as signal processing and database management will be improved by the year 2000 by factors of two or three orders of magnitude using parallel and distributed architectures. An example of the increase of bit and chip density is given in Figure 3-6. See also Mayers (1991).

3.7 Processors

The trends in processor speed [measured in Millions of Instructions Per Second (MIPS)] and fiber transmission rates, measured in Gigabits per second (Gb/s), are given by Fraser (1991) and illustrated in Figure 3-7. The increasing speed of these two technologies versus time is not very different except for the reduced instruction set computers (RISC) which are increasing at a faster rate. If this trend continues as indicated on the figure, one can expect microprocessing speeds on the order of 1 billion instructions per second (BIPS) and transmission rates around 300 Gb/s by the year 2000.

Most computers have a single processor that does the computational work such as addition, multiplication, and number comparisons. Human programmers divided the computational tasks into a sequence of steps which the processor executes one step at a time, an inherently slow process. By linking many processors together to form parallel computers, the processing speed is greatly increased.

Parallel computer architectures operating with a distributed operating system that runs on multiple, independent, central processing units transparent to the user will play an important role in the future. According to Kleinrock (1985) distributed systems can provide the processing

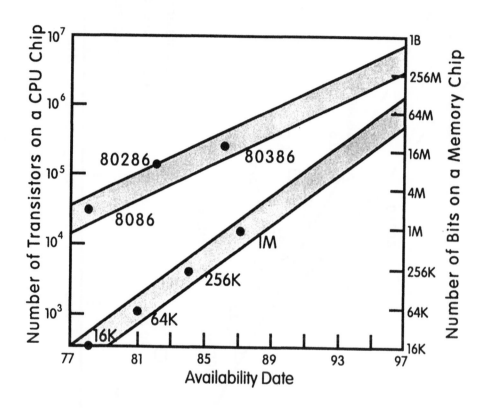

Figure 3-6. Increases in bit and chip density (Gould et al., 1991).

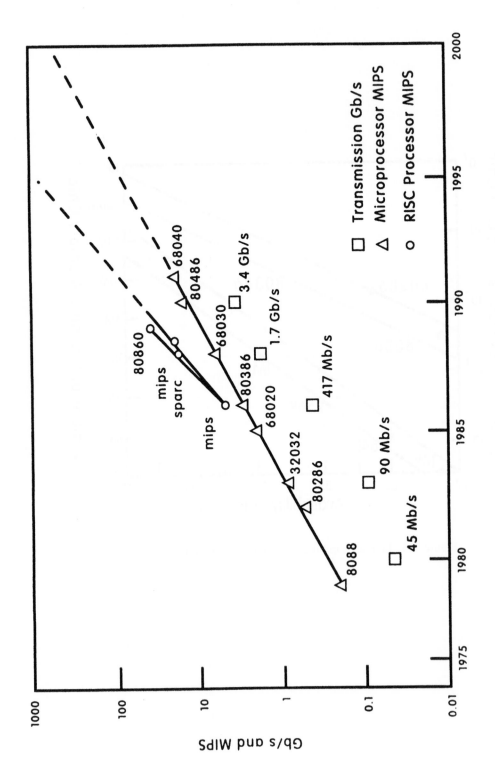

Figure 3-7. Processor and fiber transmission speeds (Fraser, 1991).

power needed to meet future user demands. Distributed data bases, distributed processing and distributed communication networks give rise to some complex architectures including parallel processing systems.

Sequential computing used in the past is being replaced by parallel computing due to advances in VLSI, parallel software, and communication technologies. Networked workstations connected by optical fiber links and optical switching may soon provide the potential for parallel computing with a great deal of processing power (Chandy and Kesselman, 1991). Massively parallel computers (MPC) imply a large number (~ 1,000 or more) processors implemented with single instruction multiple data stream. These concepts are described by several papers in a special issue of the IEEE Proceedings in April 1992. Since the number of processing elements is high and parallelism is exploited at a very fine grain, the interconnecting network plays an important role for MPCs.

3.8 Evolutionary Trends

Figure 3-7 showed a curve portraying the history and projection of computer processing unit speeds. This curve is repeated in Figure 3-8 and compared with the trend in transmission speeds for data communications over LANs and WANs.

Increasing the transmission rate of modern networks into the gigabit range introduces a new level of complexity because the delays are dominated by propagation delay. Communications techniques and protocols used for networks operating at Mb/s may be inefficient or even ineffective for networks operating at Gb/s. This latency-versus-bandwidth tradeoff is discussed in detail by Kleinrock (1992). Kleinrock concludes that due to the finite speed of light in Gb/s networks, the propagation delay for long links is much larger than the time required to transmit blocks or packets of data into the link. This introduces entirely different issues concerning flow control, buffering, error correction, and congestion control that are still unknown. Fraser (1991) notes that the ratio of backbone speed to local area network speed is expected to be 50:1 by 1997 and continue to increase.

The transmission rate for LANs and WANs may continue to rise as shown in Figure 3-8 and could approach 1 to 10 Gb/s by the year 2000. An overview of Gb/s LANs from a systems perspective is given by Kung (1992). Important applications for these high-speed networks include three-dimensional imaging and computing. More important than the immediate

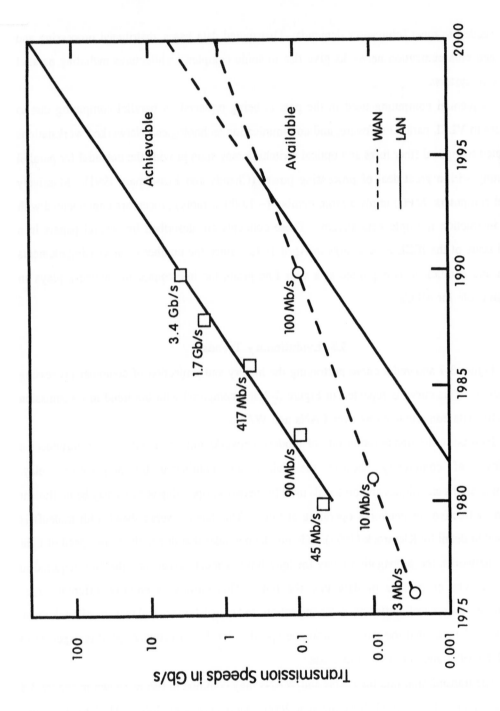

Figure 3-8. Transmission speeds for data communication (Fraser, 1991).

application is that these gigabit networks can change computing and communications in fundamental ways that have yet to be explored.

Table 3-6 summarizes the evolution of switching and multiplexing technologies beginning with conventional systems, from analog to digital, through packet and fast packet switching, and culminating in broadband ultrafast packet switching based on cell relay concepts. Figure 3-9 illustrates, in another way, the progress that has been achieved in processing and transmission technologies since 1985. The figure includes projections beyond the year 2000 when transmission speeds at terabits per second (10^{12} b/s) are potentially feasible.

This page has been intentionally left blank.

Table 3-6. Evolution of Switching/Multiplexing Technologies

	Conventional					Fast Packet					Ultrafast Packet	
Type Service	Analog Voice/Data	Analog Voice/Data	Digital Voice/Data	Low-Speed Data	Video	High-Speed Data		High-Speed Trunking	High-Speed Switch Access	High-Speed Data	Digital Broadband	Digital Broadband
Switch Technologies	Mechanical Circuit Switch	Electronic Circuit Switch	Circuit Switch	Packet	Channel	Frame Relay		Cell Relay	Cell Relay	Cell Relay	Cell Relay	Photonics
Control Technologies	Manual and Mechanical	Software	Software	Software	Software	Software		Software	Software	Software	Software	Software
Switching and Multiplexing Rate	1.2 kb/s to 9.6 kb/s	9.6 kb/s	64 kb/s to 1.5 Mb/s	56 kb/s	1.5 Mb/s to 45 Mb/s	Below 1.5 Mb/s	Below 45 Mb/s	Below 45 Mb/s	Below 45 Mb/s	Below 150 Mb/s	Below 600 Mb/s	Above 1 Gb/s
Transmission Rate	4 kHz Voice Channel	4 kHz Voice Channel	DS-0 (64 kb/s)	DS-0	DS-1 and DS-3	DS-1 (1.5 Mb/s)	DS-3 (45 Mb/s)	DS-3	DS-3	OC-3	OC-12	OC-3 and Higher
Interface	Data Modem	Data Modem	Voice Codec	Voice Codec	N/A	N/A		N/A	N/A	N/A	N/A	Optical
Physical Transmission	Copper, Radio	Copper, Radio	Copper, Radio	Copper, Radio	Copper, Fiber, Radio	Copper, Fiber, Radio		Copper, Fiber, Radio	Fiber	Fiber	Fiber	Fiber
Application	Pt-Pt Voice and Low-Speed Data	Pt-Pt Voice and Low-Speed Data	Pt-Pt Voice/Data	Pt-Pt	Broadcast and Pt-Pt	MANs		LANs and MANs	MANs and WANs	WANs	WANs	WANs
Examples	PSTN Dial-up Data	PSTN Dial-up Data	PSTN Dial-up Data	PDN and PLN	Conferencing TV	LAN Interconnect		Internet B-ISDN	SMDS	ATM/SONET B-ISDN	HDTV	Future Multimedia, Video
Availability	Disappearing Fast <5% Today	Today 50%	Today 50%	Today	Today	~1994	Today	Today	~1993	~1995	~1995	~2000+

This page has been intentionally left blank.

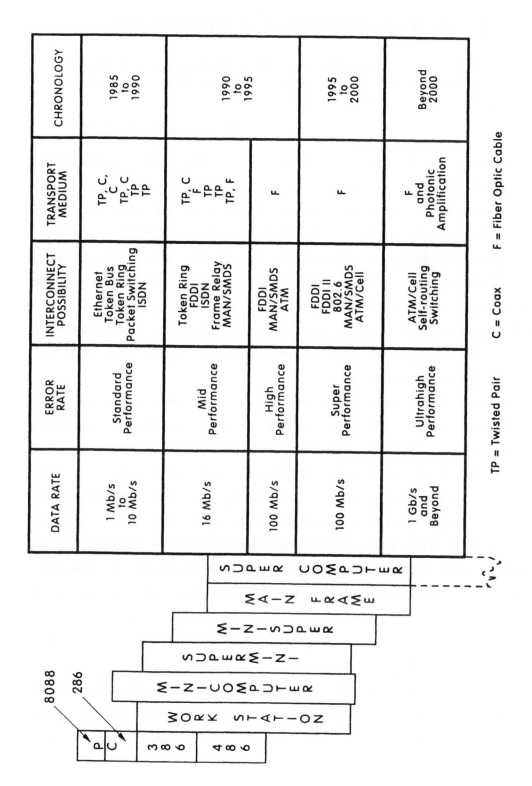

Figure 3-9. Processing and transport mechanisms.

4. The Role of Government and Standards

"To make available, as far as possible, to all the people of the United States, a rapid, efficient nationwide and worldwide wire and radio service with adequate facilities at reasonable charges."

Communications Act of 1934

Empowered by the Communications Act of 1934, the Federal Communications Commission (FCC) set out to achieve the mission defined above. See Communications Act (1934). This worthwhile goal had, in large measure, been achieved by the 1960's. In fact, by the mid 1960's, the depth of penetration of plain old telephone service (POTS) far exceeded what was originally envisioned by the sponsors of the Act. In order to achieve this so-called, "universal" and "affordable" service, the FCC initially believed that the public would best be served by a monopoly where economies-of-scale would provide the affordable part and rate averaging the universal part. About this same time, a new product was developing that would have far-reaching effects, not only on the telecommunications environment, but on the FCC as well--namely the programmable computer.

By the mid 1970's, the distinction between computer processing and communications became blurred as these two technologies converged. It was apparent that any regulation based on the dichotomy between computer processing and communications could not long endure. Over the next two decades, the FCC conducted a series of inquiries known as Computer I, II, and III. See FCC (1970, 1973, 1977, and 1986). As these inquiries progressed, the philosophy of the FCC, the Congress, and the Justice Department changed. Rather than regulate and monopolize, this new philosophy encouraged deregulation and competition. Depth of penetration of POTS would be supplemented with a new goal--breadth of services. The competitive environment under marketplace control would yield new innovative features and functions to meet the service demands of an emerging new information society. The old objective of universal, affordable POTS was not replaced, but a new objective was added, namely Peculiar and Novel Services or PANS.

This new philosophy resulted in the deregulation of customer premises equipment and enhanced services in 1981, the divestiture of AT&T in 1984, and the yet-to-be implemented open network architecture (ONA) concept in 1987. However, these changes have not come about without problems. Although many expected benefits occurred, new issues arose. How these issues evolved, and their current status, is the subject of the following paragraphs.

After two decades of controversy over competition in the carrier industry, a dramatic organizational change occurred on January 1, 1984. The year before, Judge Greene had approved the Modified Final Judgment (MFJ) dissolving the Bell system. See Green (1983). This divestiture divided what was then the largest corporation in the world with some $150 billion in assets serving over 100 million subscribers into eight independent companies, seven Regional Bell Operating Companies (RBOCs) and AT&T. Figure 4-1 indicates the geographic area covered by the seven RBOCs, and also indicates the 1991 ISDN deployment level in terms of percentage of total lines for each RBOC. This event culminated a series of deregulatory, pro-competitive, initiatives involving all three branches of the United States Government but still left the local exchange carriers as virtual monopolies. Technologies such as wireless communications systems, cable TV, and satellites could be used to by-pass the local exchange, and introduce competition in that area.

The Federal Communications Commission, the Department of Justice, the National Telecommunications and Information Administration, and Judge Greene's District Court are still (in 1993) clarifying and refining the organizational restructures imposed by the (1984) divestiture. In order to understand how these government policies and actions have affected today's networks, and may impact future structures, it is necessary to review the major regulatory events of the past, today's regulatory posture, and what may happen in the future.

Table 4-1 lists, in chronological order, several major actions and events that have occurred as the result of Government actions affecting the telecommunications industry.

One change occurred in March 1988 when Judge Greene issued his decision allowing Bell operating companies (BOCs) into the voice-mail and electronic mail markets. The decision also permits transmission of information services, but continued the ban on origination of information. Recently (1992), Judge Greene lifted the ban and allowed the BOCs to provide information gateway services and the FCC permitted the telephone industry to provide TV services.

A report and order issued by the Federal Communications Commission (FCC) following the Third Computer Inquiry (FCC, 1986) replaced structural separation requirements for enhanced services operations of AT&T and the BOCs with nonstructural safeguards. Initially this included the imposition of Comparably Efficient Interconnection (CEI) and Open Network Architecture

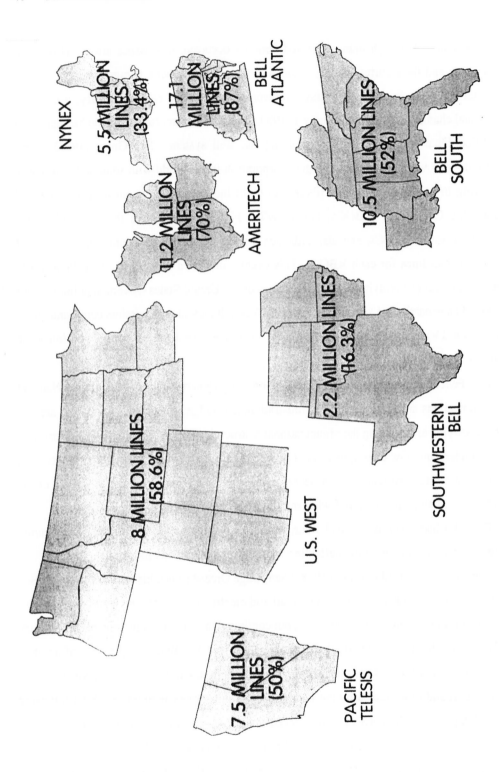

Figure 4-1. ROBC development of ISDN in terms of access lines.

Table 4-1. Government Actions Affecting the Telecommunications Industry

Year	Action	Comment
1893	Bell Patents Expired	Competition Begins
1910	Mann-Elkins Act	Interstate Commerce Regulation
1913	Kingsbury Commitment	Interconnections Required
1927	Radio Act	
1934	Communications	Enabling the FCC
1949	AT&T Antitrust Suit	Ultimately led to Divestiture
1956	Hush-a-Phone Decision	
1956	Consent Decree	AT&T out of Processing Business
1962	Communication Satellite Act	
1968	Carterphone Decision	Interconnect allowed CPE Industry Starts
1969	MCI Application Approved	Start of Long Distance Carrier Competition
1972	Domestic Satellite Decision	Open Sales Policy
1971	Computer I Final Decision	Open Field to Specialized Carriers
1974	AT&T Antitrust Suit	
1975	FCC Equipment Registration	Interconnect Market Expands
1977	Execunet Decision	
1980	Computer II Final Decision	Basic/Enhanced Dichotomy
1983	Modified Final Judgment	Divestiture with Business Restrictions on BOCs
1984	Divestiture (AT&T and BOCs)	Equal Access Required
1986	Computer III Report and Order	Open Network Architecture
1987	Computer III Supplementary Order	
1988	ONA Plans Approved in Part	
1989	BOCs Allowed in Voice and Electronic Mail Market	BOCs Permit Access to Enhanced Service Providers
1992	BOCs Allowed to Provide Information Services	Business Restrictions on BOCs Partially Lifted
1992	FCC Permits Telecos in TV Service Market	Introduced Competition to Cable TV Industry

(ONA) plans. Approval of an ONA plan was contingent on the unbundling, identification, and offering of Basic Service Elements (BSEs). The development of these ONA plans involved a complex interplay between a number of conflicting interests--the carriers, the equipment manufacturers, new entrepreneurs, regulators, and users. This complex process has been long and involved, but ultimately ONA is expected to have a major impact on the telecommunications industry and the network infrastructure in the United States. Compliance with ONA is a condition for removal of the RBOCs structural separation requirements imposed by the FCC.

4.1 The ONA Model

The Bell Operating Companies have developed a common ONA model to facilitate the development of their ONA plans. This common model, known as the Bellcore model, is illustrated in Figure 4-2. It depicts the generic elements of any ONA connection to a BOC's network.

The Bellcore model encompasses three main components: (a) Basic Serving Arrangements (BSAs), (b) Basic Service Elements (BSEs), and (c) End-user Complementary Network Services (CNSs). Under such an approach, a BSA--comprising the Enhanced Service Provider's (ESP's) access arrangement to a BOC network--must be taken as a precondition to ordering various optional BSEs associated with a particular BSA.

Despite the reliance on a common model, the plans of each RBOC still varied widely with respect to the number of specific BSAs identified and the number of BSEs offered.

Figure 4-3 lists a number of applications that the Enhanced Service Providers contemplate under ONA. These applications have been divided into five major service categories--passive, interactive, transitional, messaging, and polling. Obviously all of these would not be required in every access area, but subsets may be useful in many areas. The ONA plans were approved in part in 1988. Court proceedings have delayed the process, but ONA is expected to influence the network infrastructure for sometime to come. Standards, both national and international, will also have major impact on this infrastructure as discussed below. The material in Section 4.2 is based on a similar discussion given by Jennings et al., (1993).

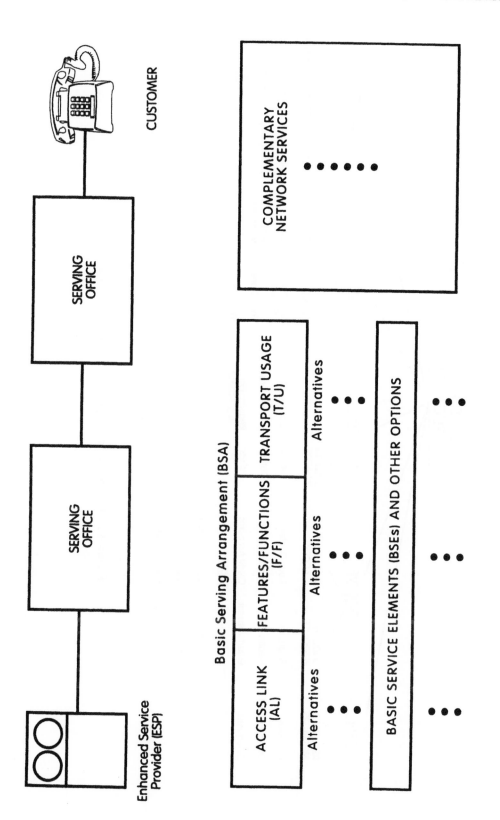

Figure 4-2. Bellcore's ONA model.

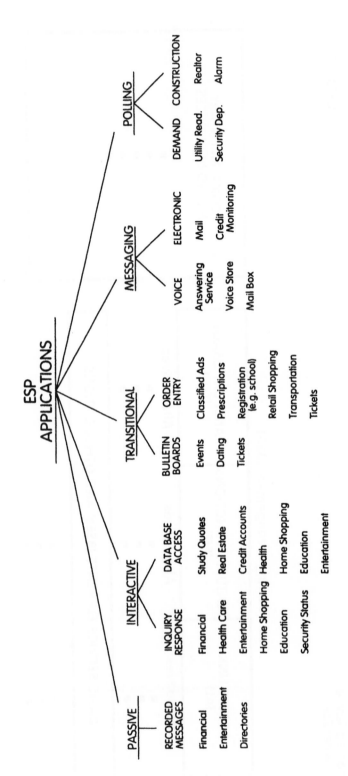

Figure 4-3.   Enhanced service providers application breakdown.

4.2 The Standard Making Process

Before describing the process for developing standards, it is useful to define what is meant by a "standard" and who needs it. Cargill (1989) defines a standard as follows: "A standard is the deliberate acceptance by a group of people, having common interests or backgrounds, of a quantifiable metric that influences their behavior and activities by permitting a common interchange."

For telecommunication standards there appear to be two viewpoints: one technical and the other functional. Technically, two pieces of equipment are standardized if they can communicate with each other or if they can both be used with the same interconnection. The alternative functional view is the documented standard that specifies approved means of accomplishing a set of tasks or functions. In this case, different implementations may meet the standard but may still be incompatible with each other due to various options.

Benefits of telecommunication and information-processing standards are often market driven. These benefits include interchangability, convenience, risk reduction, interconnectibility, safety, ease of use, and technical integration.

The development of standards is a multistep process. One typical example is shown in Figure 4-4. An estimated time scale for major processes is given on the left side of the figure and potential organizations that could be involved with each step are listed on the right. The process begins with the establishment of a need or requirement. This could come from a variety of sources including service providers, equipment suppliers, and the users themselves. Each group may approach this need from a different perspective. The providers, for example, tend to view their networks as all encompassing, capable of meeting a variety of users needs, and having long productive lifetimes. The users on the other hand are more interested in an immediate implementation to meet their specific application. Needs may also evolve from special groups formed for that purpose. For example, the International Federation for Information Processing (IFIP) tends to be a prestandards organization that investigates only the need for standards, not their development.

After the need is established, the next step is to develop a basic framework for standards development. This framework scopes out the standardization activities needed to develop a particular standard or set of standards, e.g., for network management. This framework provides an overview of what is, and what is not, to be standardized. Detailed models then refine the

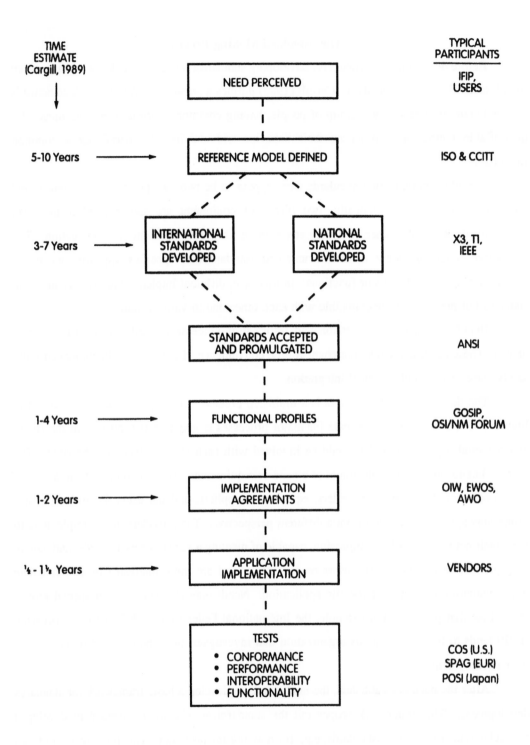

Figure 4-4. The standards-making process.

basic framework. This functional architectural model leads to standards development by national and international bodies. These bodies typically concentrate on standards for specific environments such as local area networks, or long-haul networks. Some are concerned with terminal access to transmission systems, some for computer communications, others for Integrated Services Digital Networks (ISDNs) or for telephony. The ultimate goal of these standards is to enable the development of interoperable, multivendor products for information processing systems and telecommunication networks.

Once the standards are developed, accepted, and promulgated by industry providers, other user-oriented organizations must develop specifications which identify the options and sets of protocols called "profiles" or "suites" that a given implementation should support. Separate functional profiles may be needed for different applications (e.g., electronic mail, file transfer, or network management) and for different physical networks (e.g., connection-oriented or connectionless). Thus, the Government's Open System Interconnection Profile (GOSIP) defines Federal procurement profiles for "open system" computer network products. Such profiles may change as technology improves and as standards evolve. New profiles are added as new applications arise.

The functional profile specifies what sets of functions are to be implemented and how they should appear to external systems. There are many possible ways to implement a profile in hardware and software, but, externally, the functions should all appear identical.

Profiles may be derived from many sources and various architectures. Some vendors have profiles based on their proprietary architectures such as the Systems Network Architecture (SNA) used in IBM networks. The profile is used to provide interoperability, but interoperability still requires agreements on how they should be implemented. These so-called implementation agreements (IAs) or system profiles are derived by consensus among users, vendors, and system integrators at various forums and workshops both national and international. For example, the Open System Interconnection (OSI) Implementors Workshop (OIW) that is sponsored by NIST and the IEEE Computer Society is developing IAs for emerging network management standards. Implementors workshops including those in Europe and Asia may submit profiles to the International Standards Organization (ISO) which can issue International Standardized Profiles (ISPs).

Products implemented according to the IAs must then be tested to certify that they meet specifications. There are several kinds of testing, including

Conformance Testing to verify that an implementation acts in accordance with a particular specification (e.g., GOSIP).

Performance Testing to measure whether an implementation satisfies the performance criteria of the user.

Functional Testing to determine the extent to which an implementation meets user functional requirements.

Interoperability Testing to duplicate the "real life" environment in which an implementation will be used.

Most vendors have not yet had their equipment certified for compliance with presently established standards because testing agencies in early 1993 are still in the process of establishing criteria for compliance testing and certification. There are a number of specific national and international organizations actively working to evolve this type of testing criteria. One is the Corporation for Open Systems (COS), a U.S.-based company developing tests for the OSI Reference Model's Layers 1 through 4, which deal with physical, data link, network, and transport services and protocols. Another is the Standards Promotion and Applications Group (SPAG), a European group establishing tests for Layers 5 through 7, dealing with session, presentation, and application services and protocols. Yet another is NIST which is overseeing the setting of standards for GOSIP.

The entire process is estimated to take anywhere from 11 to 22 years, but actually it is never complete since changes occur and new standards evolve as technology and needs change. Figure 4-5 shows the major standards organizations involved in developing telecommunications, radio, and information processing standards in the United States. The Telecommunications Industry Association (TIA) formed in 1988, recently accredited by the ANSI, plays a leading role in developing standards for telecommunications equipment and systems, fiber optic components and systems, and for mobile and cellular radio equipment. Organizations such as the European Telecommunications Standards Institute (ETSI) and the Information Technology Steering

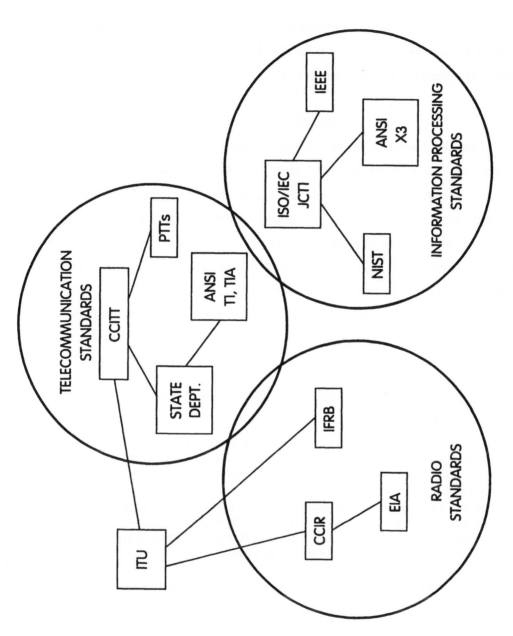

Figure 4-5. Major groups involved with standards for telecommunications and information processing.

Committee (ITSTC) are expected to develop standards for the European Community (EC) which will impact global networks. For a detailed description of the total organization structure and the standards making process see Jennings et al., (1993).

4.3 The OSI Model

An important result of the international standards-making process is the open system interconnection (OSI) reference model and the set of related standards resulting from this model. This set of international OSI standards attempts to ensure the interoperability of future telecommunications networks. They are, however, still incomplete (e.g., B-ISDN) and many organizations are attempting to resolve the problems. The OSI concept is addressed briefly because it has considerable impact on future network architectures such as B-ISDN and on the implementation of networks based on this model. See CCITT (1988c).

The seven-layer OSI model is illustrated in Figure 4-6. Narrative descriptions of the value-added services provided by protocols in each layer to the adjacent layer above are quoted from Federal Standard 1037B (1991). They are as follows:

Media Layer: Layer 0. This is not currently a Federal standard but is concerned with the infrastructure of the network.

Physical Layer: Layer 1. The lowest of seven hierarchical layers. The Physical Layer performs services requested by the Data Link Layer. The major functions and services performed by the Physical Layer are: (a) Establishment and termination of a connection to a communications medium; (b) Participation in the process whereby the communication resources are effectively shared among multiple users, e.g., contention resolution and flow control; and, (c) Conversion between the representation of digital data in user equipment and the corresponding signals transmitted over a communications channel.

Data Link Layer: Layer 2. This layer responds to service requests from the Network Layer and issues service requests to the Physical Layer. The Data Link Layer provides the functional and procedural means to transfer data between network entities and to detect and possibly correct errors that may occur in the Physical Layer.

Network Layer: Layer 3. This layer responds to service requests from the Transport Layer and issues service requests to the Data Link Layer. The Network Layer provides the functional and procedural means of transferring variable length data sequences from a source to a destination, via one or more networks while maintaining the quality of service requested by the Transport Layer. The Network

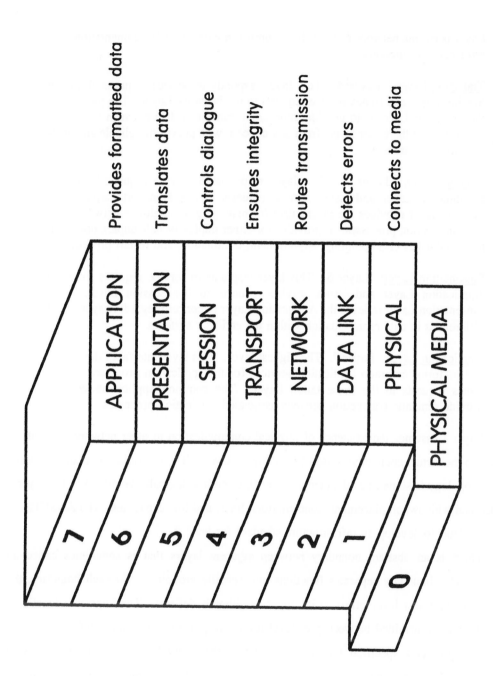

Figure 4-6. Protocol reference model for data communications.

Layer performs network routing, flow control, segmentation/desegmentation, and error control functions.

<u>Transport Layer</u>: Layer 4. This layer responds to service requests from the Session Layer and issues service requests to the Network Layer. The purpose of the Transport Layer is to provide transparent transfer of data between end users, thus relieving the upper layers from any concern with providing reliable and cost-effective data transfer.

<u>Session Layer</u>: Layer 5. This layer responds to service requests from the Presentation Layer and issues service requests to the Transport Layer. The Session Layer provides the mechanism for managing the dialogue between end-user application processes. It provides for either duplex or half-duplex operation and establishes checkpointing, adjournment, termination, and restart procedures.

<u>Presentation Layer</u>: Layer 6. This layer responds to service requests from the Application Layer and issues service requests to the Session Layer. The Presentation Layer relieves the Application Layer of concern regarding syntactical differences in data representation within the end-user systems.

<u>Application Layer</u>: Layer 7. The highest layer. This layer interfaces directly to and performs common application services for the application processes; it also issues requests to the Presentation Layer. The common application services provide semantic conversion between associated application processes.

Layer 1 assumes the existence of physical communication to other network elements as opposed to the virtual connectivity used by the higher layers. The transmission media including network topology is sometimes denoted as layer 0, since it is logically below layer 1. Layer 0 is concerned with switch placement, concentrators, lines, and line capacities. The ANSI/TIA is making a effort to develop an infrastructure standard for level 0.

There is an abstract boundary between adjacent layers that is sometimes called an interface. This boundary separates functions into specific groupings. At each boundary, the service that the lower layer offers to its upper neighbor is defined. The important functional entities that are transmitted between peer level layers are protocol data units (PDUs).

The protocols specified for all layers define the network's functional (or protocol) architecture. The specification of these protocols is needed to implement a service to an end user. Implementation of these protocols in hardware and software can be accomplished in many ways. The details of the implementation are not part of the architecture. One major advantage of this layered architecture concept is that lower-layer implementations can be replaced as technologies advance; for example, when a fiber link replaces a coaxial cable. The only requirement is that any new implementation provide the same set of services to its adjacent upper layer as before.

It is not always necessary to implement every layer or every protocol within a layer. For example, error checking, a function of Layer 2, may not be necessary on links with low error characteristics.

There are limitations to the OSI model. For example, it may be difficult to apply to certain distributed systems where computing functions are dispersed among many physical computing elements. The model does not, in its present form, represent important existing and future services such as telephony. It tends to restrict certain functions to end systems. This can be inconvenient where such functions could be better performed by the network itself.

Figure 4-7 illustrates the application of this model for connecting a user to a computer program via two intermediate packet switching nodes. Note that only the lower layers 1 through 3 are involved at an intermediate node. Layer 4 is concerned with the end-to-end integrity of the information transferred between systems A and B. Actually, a functional architecture based on this model and the subsequent implementations could be different for each link in this configuration. Thus, the protocols from System A to the first node may be entirely different than protocols 1 through 3 between the switching nodes. Even the physical transmission media may differ. Figures 4-8a and 4-8b indicate the relationships between the OSI protocol reference model and conventional data terminal equipment (DTE), as well as data communication equipment (DCE). Figure 4-8c relates this reference model to the functional grouping of the elements in an ISDN. Figure 4-8d illustrates one implementation of the two communicating systems with a subnet containing the two nodal switches. It is also possible for the reference model to take on more dimensions to include the network management and control functions.

Based on the OSI reference model, it is possible to define a number of different functional architectures for a given end-service. This is accomplished by selecting appropriate protocols for each of the seven levels (Linfield, 1990). An example of protocol stacks leading from two applications (file transfer and electronic mail), through the seven OSI levels to seven different physical transmission media is shown in Figure 4-9.

A similar protocol stack has been specified for government use by the National Institute of Standards and Technology (NIST) and is called the Government OSI Profile (GOSIP). Since

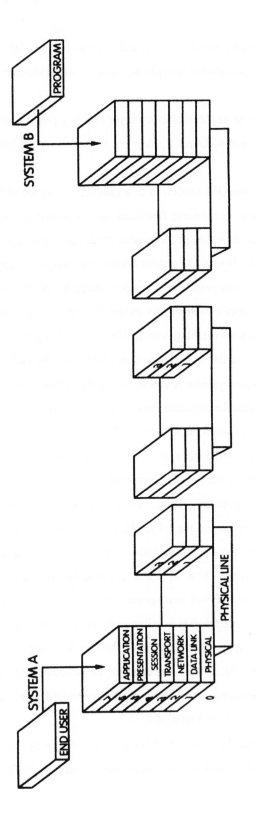

Figure 4-7.   Application of protocol reference model to a network with intermediate nodes.

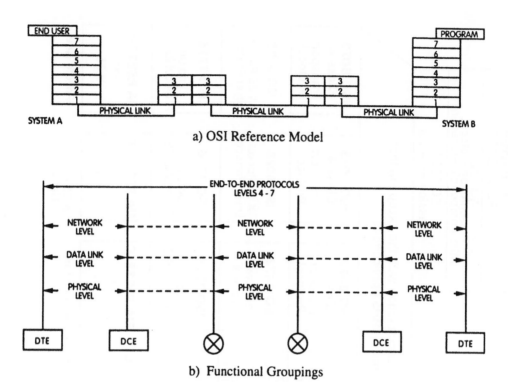

a) OSI Reference Model

b) Functional Groupings

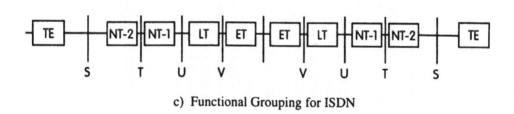

c) Functional Grouping for ISDN

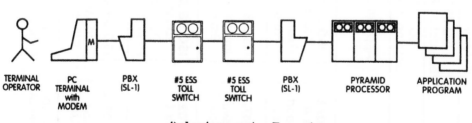

d) Implementation Example

Figure 4-8. Architectural models and an implementation.

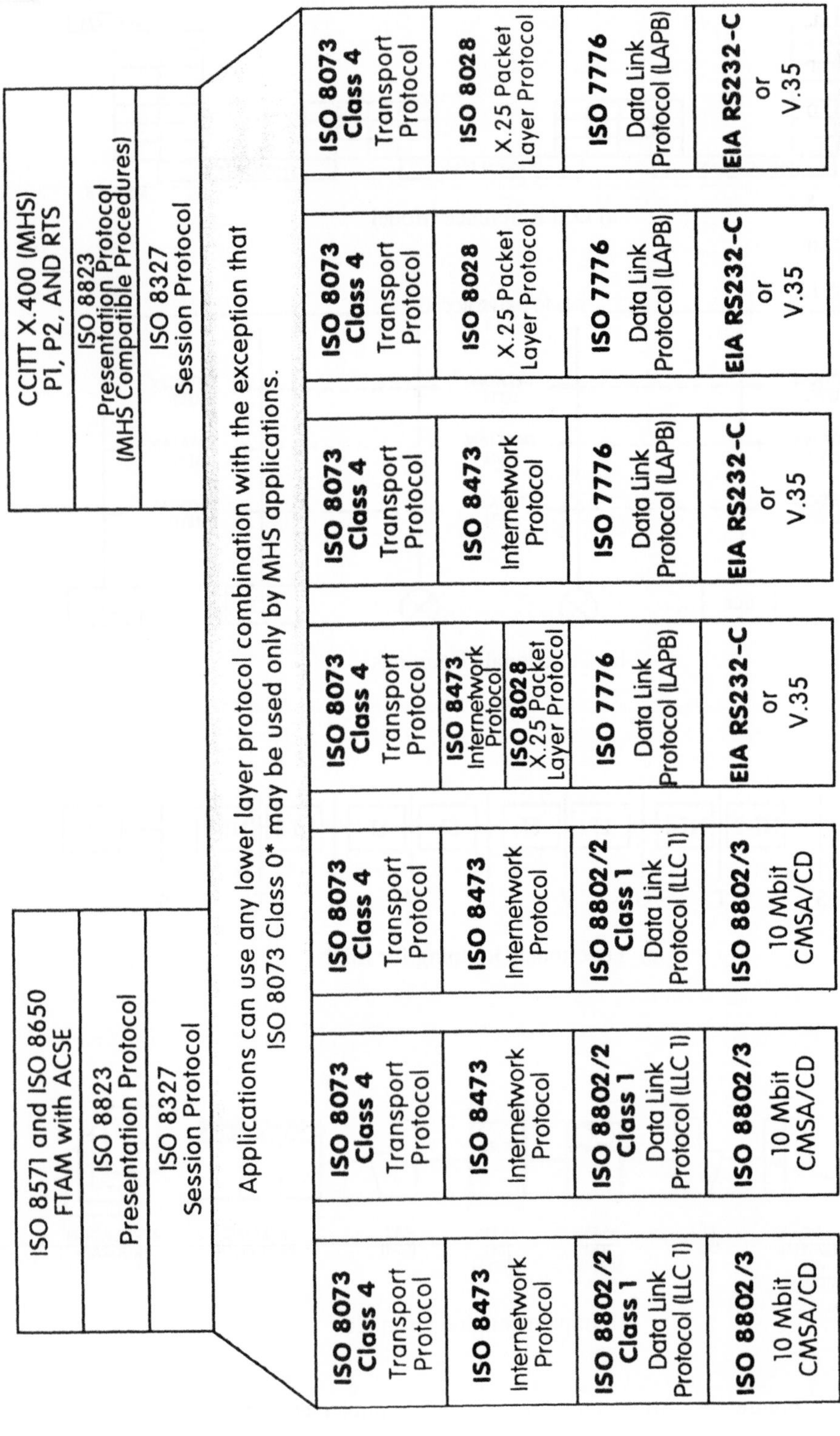

Figure 4-9. Protocol combinations defining specific network architectures.

1990, GOSIP has been mandated as a Federal Information Processing Standard and is defined in Federal Information Processing Standard (FIPS) Publication 146. All government agencies must conform to GOSIP in any future procurement of network products. GOSIP is updated each year. Version 2 became mandatory in August 1991. Version 1 applications covered File Transfer, Access and Management (FTAM), Message Handling Systems (MHS), and the Association Control Service Element (ACSE). Version 2 added applications for virtual terminals and document interchange. Version 2 will support X.25 interfaces, and IEEE 802.2 to 802.6 LAN networks, as well as ISDN. Version 3, effective in 1993, will add network management, directory services, and will support the Fiber Distributed Data Interface (FDDI).

5. The Impact of Market Forces and User Needs

"You can't always get what you want, but if you try, sometimes you get what you need."

The Rolling Stones, 1972

It is important here to distinguish between network users, providers, and suppliers. User demands for service influence the service providers. The providers' needs in turn influence the equipment manufacturers and suppliers. Other forces may be acting on all three. Government policies and standards development affect the providers, and to some extent the suppliers. See Figure 5-1.

Providers, suppliers, and users all have different goals and perspectives of their telecommunications world. For example, the provider, for economic reasons, develops a single network that he hopes will meet the needs of many users. The users, on the other hand, want optimum choices for specific applications.

Traditionally the telecommunications provider conducted basic research, developed equipment and facilities, and applied them to meet operating needs but, not always the users' needs. Today, the user is becoming more involved in the entire process, and the suppliers more independent. Table 5-1 summarizes today's different perspectives between users, providers, and suppliers. Their conflicting interests are discussed in the following two subsections.

Table 5-1. User, Provider, and Supplier Perspectives

Users	Providers	Suppliers
price terminal choices service reliability service selections speed of service quality of service time to implement	cost to implement efficiency survivability time between failures traffic types traffic volume	cost to manufacture technical feasibility proprietary market innovativeness maintainability

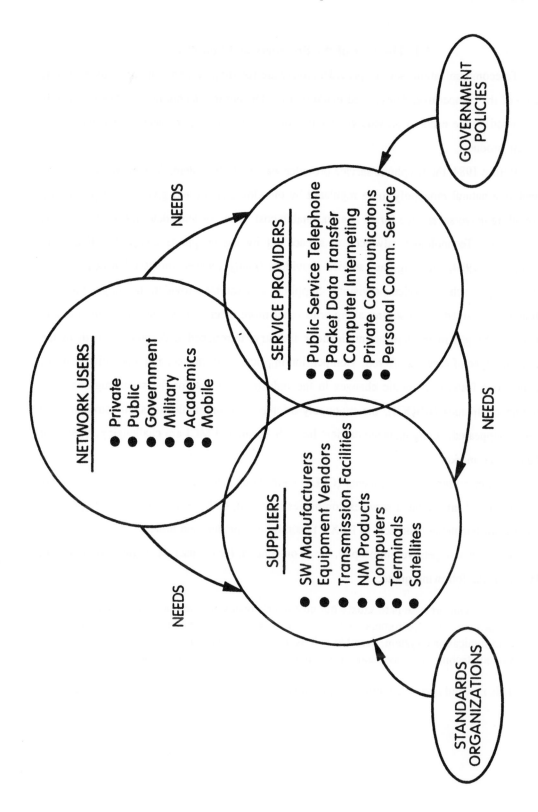

Figure 5-1. Major participants in telecommunications infrastructure.

5.1 The Role of the Providers and Suppliers

Telecommunications service providers today are focusing more and more on users' needs because of the competitive, deregulated marketplace. The entire telecommunications industry is being forced by the users to provide greater flexibility, control, responsiveness, and integration into their products.

Before 1984, the telephone service provider was the Bell System. Telephone service was viewed as a natural monopoly to be regulated by the Government and given the responsibility of providing universal service for all at a reasonable cost. The "system" determined users' needs not the user. Technologies advanced at a pace set by the telephone company. Following divestiture in 1984, the long-haul service providers finally focused on users' needs using a marketing approach to service provision. Suppliers and service providers no longer took the customer for granted. Divestiture, along with other more recent technological, economic, and regulatory changes in the telecommunications industry, introduced widespread growth in the number and type of services offered. Competition in price, performance, and innovative features and functions fostered dramatic changes in the industry. Except for the seven regional Bell operating companies (RBOCs), the monopoly-oriented telecommunications infrastructure has largely disappeared. Long distance carriers like US Sprint and MCI are challenging AT&T for a share of the market.

Customer premises equipments (CPEs) were deregulated almost 25 years ago. Today several thousand interconnect services companies buy, sell, or lease a wide variety of products ranging from telephone hand sets to PBXs, facsimile equipment, and more.

This new competitive philosophy, based on concern for the user, is summed up by Fluent (1987) with the following quote;

> "Our premise is to satisfy our customer's needs, and we have an obligation to provide our customers with the most efficient, streamlined, up-to-date communications system possible. It is important therefore, to investigate the various technologies available to see if they will work for our customers."

This new industry-wide approach puts a new responsibility on the user, as described next.

5.2 The Role of the User

Users today play a more active role in developing new applications, new standards, and even in formulating new regulations. This new involvement, following divestiture and the subsequent deregulation of the industry, resulted in a more competitive environment, gave users a greater influence over what equipment is produced, and increased the variety of services that were offered. At the same time, users found a pressing need to manage their own networks and to select special features and functions which met their business requirements for maximum efficiency at the lowest cost. Instead of buying just what is available, users today buy what they need, or have it developed. Because of this, user devices and features have changed substantially in the past few years. End user premises equipments, terminals such as personal computers (PCs), local area networks (LANs), facsimile (FAX), and multimedia workstations have all grown dramatically in the last few years as shown in Figure 5-2.

The ISDN evolution is another example of the user's role in the changing infrastructure. Users have been actively involved in defining their needs and determining where new technologies can best serve these needs.

Finally, users are becoming actively involved in standards organizations and the standards setting process. Users want standards that ensure equipment compatibility from the multivendor sources. Ubiquitous technical standards and availability of services are obviously to the users' advantage. It is also to the users' advantage to promote competition throughout the industry by reducing barriers to entry and by participating in user organizations as well as developing standards.

Users today have considerable impact on the telecommunications infrastructure. For example, they now have the ability to by-pass the public network with private facilities including very small aperture terminals (VSATs). User influences are also already being felt in office automation. In the traditional office, the telephones and the PABX that controlled them were the most dominant fixtures. Records and correspondence were produced by typewriters, and later by shared wordprocessors. Storage facilities included file cabinets and punched cards. In the near future, offices may use multimedia workstations with multitasking, interconnected with LANs, MANs, WANs, using digital PABXs or wireless media, and ISDNs. Spreadsheets, desktop publishing, graphic art, electronic mail, and voice messaging are in demand already in

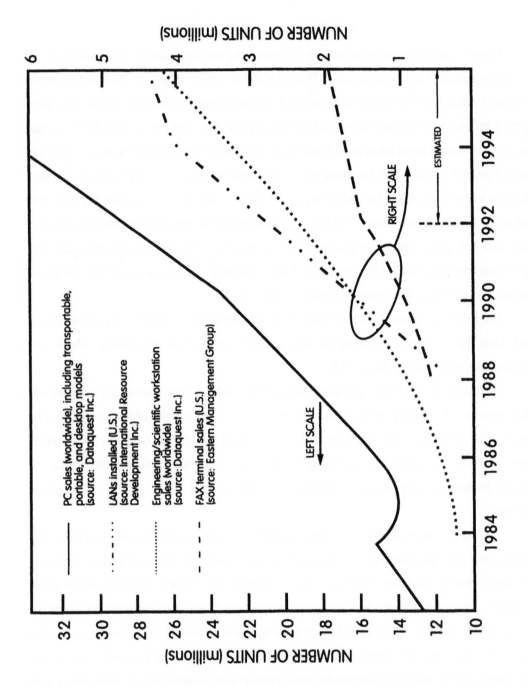

Figure 5-2. Growth in network-attached user devices (Anonymous, 1990).

many offices. Video teleconferencing, cordless phones, wireless LANs, and expert systems for network management could become commonplace.

The information business involves a complex interplay between public and private institutions, between government and industry, and between national and international entities. The participants in the information business are concerned with the acquisition, packaging, processing, storage, transmittal, and distribution of all kinds of information including voice, data, imagery, and video. They are constantly striving to make business more profitable by adding more features, operating more efficiently, and exchanging information faster. This is the purpose of many of the emerging new products, systems, and services as described in the next section.

6. The Emergence of New Products, Systems, and Services

"There is nothing more difficult to plan, more doubtful of success nor more dangerous to manage than the creation of a new order of things."

Machiavelli, 1513 A.D.

The following subsections discuss a number of new products, systems, and services including multimedia communications, electronic messaging, and video services. Multimedia communications is the field referring to the representation, storage, retrieval, and dissemination of machine-processible information expressed in multimedia, such as text, voice, graphics, images, audio, and video. Electronic messaging covers a broad field ranging from telex and facsimile to voice mail and electronic data interchange (EDI). Video systems and services include video telephony, multicast video teleconferencing, compact video disk storage, videotext, and the like. The products, systems, and services that will actually dominate in the future are difficult to assess as noted by Machiavelli almost 500 years ago! The information services market is expected to grow since RBOCs are now able to market their own information services according to Newsfront (1991). Table 6-1 depicts this expected growth through 1994.

6.1 Advanced Terminals and Services

The advent of high-capacity storage devices, economical but powerful workstations, and high-speed integrated-services digital networks has led to a variety of multimedia services using broadband terminals. Multimedia documents, when communicated with short delays, have applications in medicine, education, scientific research, travel, real estate, banking, insurance, advertising, and publishing. Future transactions could be conducted by groups using computer-controlled cooperative workstations exchanging multimedia documents. Publicly available information services from multimedia, multilocation data could also expand according to Irven et al., (1988). Multimedia software demonstrated at trade shows combining TV, compact disk (CD), and PC technologies in a convenient service package could revolutionize the entertainment and business field. These service packages are used to create multimedia presentations combining music, voice, animation, text, photo, and video images on optical disks. Compact disk

Table 6-1. Information Service Markets in the United States in Billions of Dollars (estimated by Link Resources, Inc.)

	1989	1990	1991	Projected 1992	Projected 1993	Projected 1994
On-line transaction processing	$ 2.590	$ 2.753	$ 2.927	$ 3.483	$ 4.120	$ 4.379
Alarm monitoring/telemetry	2.176	2.502	2.827	3.166	3.544	3.969
Telemessaging services	1.025	1.096	1.172	1.279	1.369	1.482
Voice messaging	0.157	0.220	0.282	0.367	0.489	0.666
Electronic messaging	0.464	0.580	0.737	0.958	1.274	1.707
Database services	8.587	9.675	10.916	12.336	13.962	15.829
Residential data services	0.235	0.272	0.319	0.373	0.434	0.505
Voice information services	0.726	1.048	1.342	1.609	1.879	2.113
Enhanced facsimilie	0.020	0.045	0.059	0.078	0.104	0.135
Electronic data interchange	0.097	0.160	0.264	0.435	0.696	1.114
Value-added network services	0.724	0.790	0.861	0.935	1.018	1.104
Business video services	0.066	0.078	0.092	0.112	0.128	0.143
Total	$16.867	$19.219	$21.798	$25.131	$29.017	$33.146

read-only-memory (CD-ROM) systems store over 500 Mbytes of multimedia information for playback on the PC. This information can be accessed quickly and manipulated much faster than video tapes, and with better quality.

Electronic messaging is a fast-emerging technology. For example, it is predicted that by the year 2000 over 40 million people in the United States will be active users of electronic mail (Figure 6-1).

Electronic data interchange (EDI) is another important trend. EDI refers to the computer-to-computer transmission of business documents using standard formats that are readable by the computer. This automated process is already being used by the transportation, retail, and health industries to improve the efficiency of doing business.

Videotex services, whereby users can access sources of information for a variety of purposes, are available. Examples include Prodigy, Compuserv, and Delphi. These three alone serve over 2 million customers.

Audiotex systems that permit telephone callers to access prerecorded information are proliferating. Voice response systems that allow callers to interact directly with a computer are also now beginning to appear.

Enhanced Service Providers (ESPs) are offering numerous applications and services designed to increase productivity and to reduce costs.

Video teleconferencing has increased in popularity due, in part, to standards that make multipoint interoperability more feasible. Picture quality can be enhanced at lower transmission rates (112 and 384 kb/s) using sophisticated coder-decoder systems. In the past, video conferences were conducted from special studios designed for that purpose. In the future, desktop video systems, some of which are already on the market, are expected to prevail. See for example, Gold (1992). These desktop systems, however, are not accessible via LANs installed over the last 10 years. Conventional LANs currently operate at approximately 10 Mb/s and this transmission rate is shared by several workstations. Full motion video requires continuous high data rate transmissions, often in real time. Video LANs of the future must be capable of providing a constant bit rate channel for each desktop workstation.

Many of the traditional enhanced services are provided by value-added networks (VANs). VANs offer enhanced data transmission services such as speed and protocol conversion along with packet switching, error detection, and correction. Services such as electronic funds transfer,

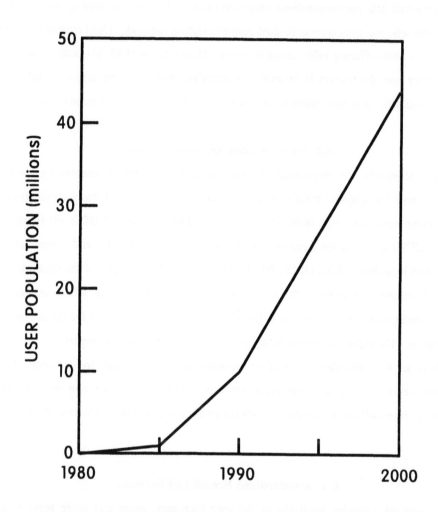

Figure 6-1. Users of electronic mail (estimate by A.D. Little, Inc.).

electronic mail, and energy management are considered VAN offerings. They fill the niche between the small data communications user who needs only a dial-up analog line and the large corporate user who has their own dedicated communication network. Until recently, the RBOCs were restricted from offering information services. However, in 1992 this restriction was lifted and their entry into the market is expected to stimulate growth in the area of residential data services such as videotext and information gateways, and in voice and electronic data messaging.

6.2 User/Provider Interconnections

User terminals for multimedia communications, electronic messaging, and video applications must interconnect locally, nationally, and globally to meet many applications. The interconnecting networks range from fiber distributed data interfaces (FDDI), distributed queue dual bus (DQDB) networks for local and metropolitan area distributions, and narrowband ISDN (N-ISDN) and broadband ISDN (B-ISDN). Figure 6-2 depicts a mixture of the technologies that interconnect information users with information providers using these advanced networking schemes. Intelligence provided by the B-ISDN allows the user to browse for the desired source of information. The networks involved are described elsewhere in this report.

There are basically three types of transmission services available to the users: dedicated lines (leased private lines), carrier services (reserved transmission time), and switched services (unreserved). Narrowband and broadband transmission types and rates are described in the next subsection.

6.3 Standardized Broadband Services

As network capacity available to the user increases, more and more services can be supported. The CCITT has proposed two initial rates for B-ISDN. One of approximately 150 Mb/s and the other approximately 600 Mb/s for single and multiple video applications. Higher rates may follow. A local exchange could handle multiple ISDN subscriber loops using twisted wire pair at 1.544 Mb/s for data and voice plus several B-ISDN subscribers using optical fiber for data, voice, and video. According to Stallings (1990), the 150 Mb/s rate seems adequate for most office subscribers-to-network directions and 600 Mb/s should suffice for the network-to-subscriber direction. Other rates including narrowband ISDN rates would also be supported, as shown in Figure 6-3.

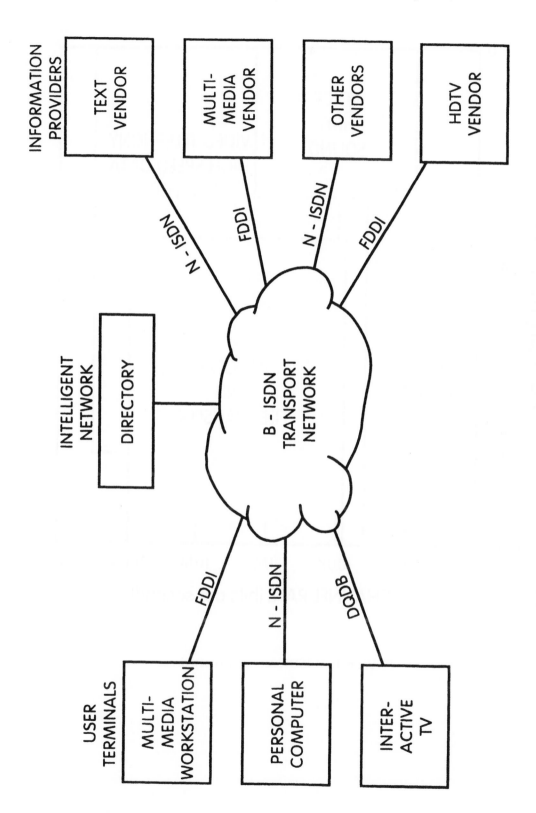

Figure 6-2.   Interconnecting information users with information providers via B-ISDN (from Irven et al., 1988).

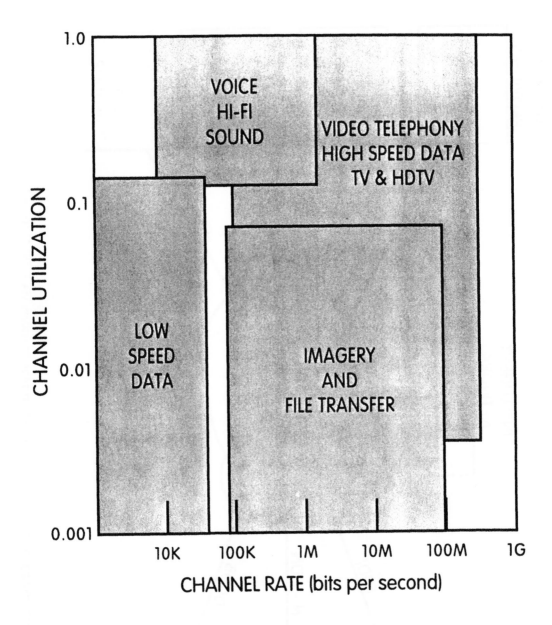

Figure 6-3. B-ISDN service characteristics.

Recommendation I.121 given in CCITT (1988a), defines the broadband aspects of ISDN and includes the service classifications. These service classifications are based on existing recommendations for ISDN and include two main categories: interactive services and distribution services, as shown in Figure 6-4. Interactive services provide a two-way exchange of information between users whereas distribution services are primarily one-way. The subcategories are defined below:

conversational services provide the means for bidirectional communications and support general transfer of data

messaging services offer communication between individual users via storage units such as store-and-forward, mailbox, and message handling functions

retrieval services provide the user the capability to retrieve information on demand from public storage centers

distributed services without user presentation control are essentially broadcast services. With presentation control, the user can control start and order of information as in teletex.

The narrowband and broadband rates proposed by the CCITT (1988a) are given below in Table 6-2. Note that the H22 and H4 rates must be multiples of 64 kb/s.

Table 6-2. Narrowband Channel Rates and Proposed Broadband Channel Rates

Narrowband		Broadband	
D	16 or 64 kb/s	H21	32.768 Mb/s
B	64 kb/s	H22	43 to 45 Mb/s
H0	382 kb/s	H4	132 to 138.24 Mb/s
H11	1.536 Mb/s		
H12	1.92 Mb/s		

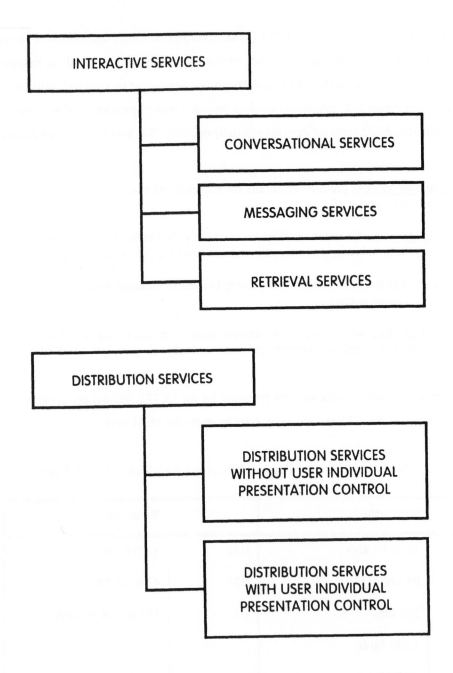

Figure 6-4. Classification of broadband services according to CCITT (1988a).

The broadband channels were selected to support the following:

- broadband unrestricted bearer services
- high-quality broadband video telephony
- high-quality broadband videoconference
- existing quality TV distribution
- high-definition TV distribution.

Only the last service would require the H4 rate. The Appendix to this report lists other possible broadband services as given in the annex to Recommendation I.121 (CCITT, 1988a). This table contains possible services, their applications, and some possible attribute values describing the main characteristics of the services.

It is expected that truly integrated multimedia services involving voice, high-speed data, image, and video will eventually emerge and penetrate the business and residential markets. These multimedia services have the potential to profoundly impact and transform those markets, and ultimately the nature of the work place.

7. The Evolution of Networks

"This planet is currently laced with many types of computer/communication networks at all levels. There are wide area networks, packet switched networks, circuit switched networks, satellite networks, packet radio networks, cellular radio networks, and more and they are mostly incompatible with each other."

L. Kleinrock (1985), Chairman of
Computer Science at UCLA

Kleinrock's seven-year-old quote still applies, except that today one should add broadband networks, personal communication networks, software-defined networks, virtual private line networks, and intelligent networks to his list and note that bridges, routers, gateways, and new standards have reduced some of the incompatibilities. However, new vendors in today's multivendor environment coupled with advancing technologies have introduced still more incompatibilities as the network infrastructure continues to evolve. The following subsections describe the evolutionary processes occurring today and the networks resulting therefrom.

7.1 Globalization of Telecommunications Service

As the world moves toward an international economy dominated by globalization of production and markets, the transport and telecommunications infrastructures will play major roles in the success of global enterprise structures. The global networks of the future can extend traditional voice and data services across national boundaries to both residential and business markets. New service opportunities such as HDTV, electronic data interchange (EDI), imaging, video conferencing, and personal communication services (PCS) could be added. PCS would offer unprecedented mobility and ubiquity by delivering a service over both wireless and wireline facilities to any subscriber, at any time, and at any place. Networks are expected to evolve gradually over time as analog and copper-based systems are replaced with digital fiber-based intelligent systems, as cells replace packets, and broadband replaces narrowband.

Major factors impacting not only this globalization process but the evolution of the entire telecommunications infrastructure are discussed below. These factors are digitization, integration, packetization, privatization, and standardization. Again, keep in mind that the speed of evolution is influenced by the installed base and the impact of amortization on the implementation of any new technologies.

7.1.1 Digitization

There are a number of advantages to digital transmission that have led to its continuous replacement for analog circuits. These advantages include improved quality particularly for long-haul circuits, reduced cost due to increased transmission efficiency, and increased security using bulk encryption techniques.

Since the AT&T divestiture in 1984, the amount of long-distance traffic has increased 2.5 times with most of that due to the addition of digital facilities. Figure 7-1 indicates the analog and digital portions of long-distance traffic in terms of billions of minutes per year for the period 1983 through 1989. Only four years ago approximately half the long-distance voice traffic was still analog.

After divestiture, the regional Bell operating companies (RBOCs) continued to digitize their offices. Figure 7-2 shows the percent of lines served by digital offices in 1991. The average approaches 50% for the RBOCs and over 85% for the independents.

The migration toward digital systems not only enhances performance due to excellent reproduction, but digital systems provide opportunities for new service features and for user participation in network management and control.

7.1.2 Integration

There are various ways that integration may take place in a network. Figure 7-3 shows how the integration process can evolve in switching and transmission. Figure 7-3a depicts the conventional analog, circuit-switched system which was designed primarily for plain old telephone service (POTS) operating over a 4 kHz bandwidth. Digital data were converted into tones that could be transmitted over 4 kHz analog channels using modems (M). Subsequently, packet switched networks evolved to carry digital data more efficiently. These packet networks permitted computers to communicate reliably at considerably less cost than the circuit-switched systems. With suitable controls to minimize delays, packet networks could carry analog voice signals with coder and decoder functions performed by codecs (C) as in Figure 7-3b.

Currently, there are still many circuit-switched public telephone networks that co-exist with separate packet-switched public data networks (PDNs) as in Figure 7-3c.

Integration of voice and data in either switching, terminals, or transmission now exists as in Figure 7-3d, e, and f. Integration of both switching and transmission facilities exists in many

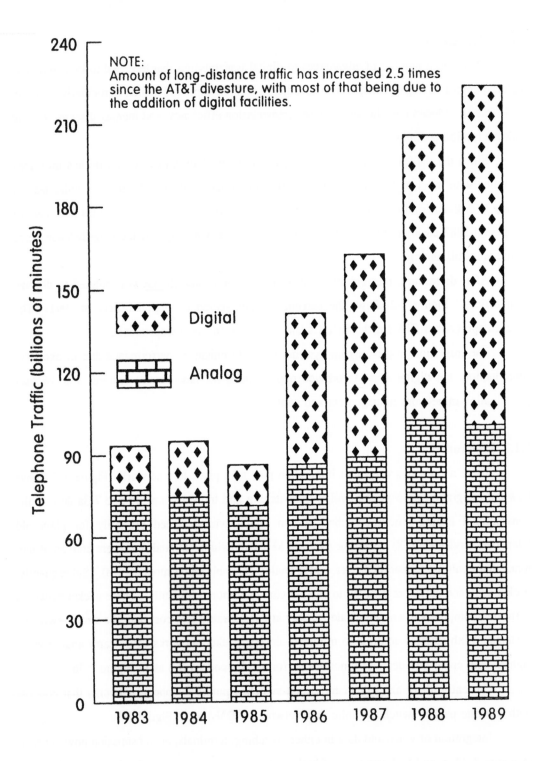

Figure 7-1. Long distance annual telephone traffic.

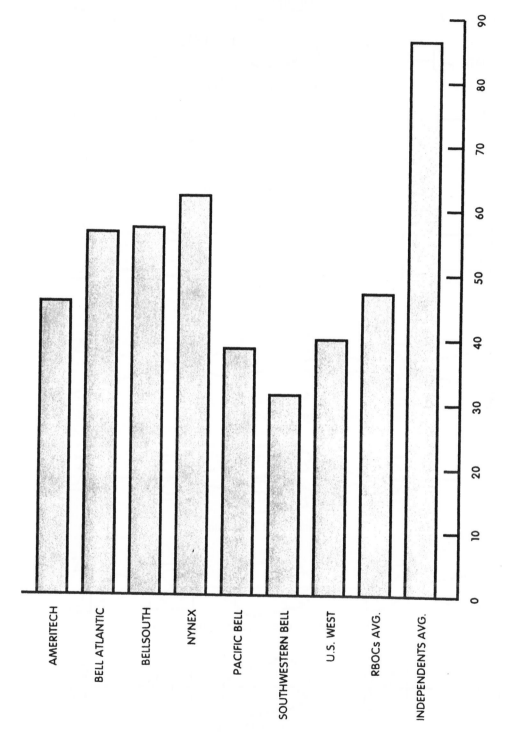

Figure 7-2. Percent of lines served by digital offices in 1991 (Smith, 1992).

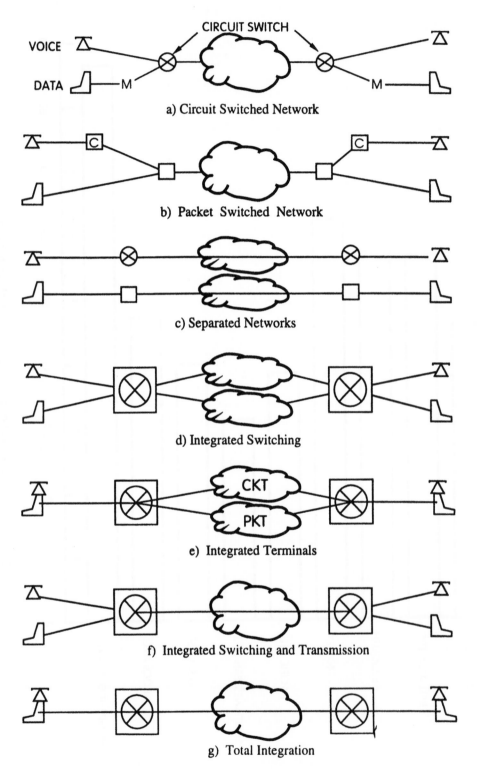

Figure 7-3. Architectural configurations and their evolutionary process.

networks (Figure 7-3f) and total integration (Figure 7-3g) on a few others. The ATM/SONET concept using cell relay provides either circuit switched or packet switched modes. Users desiring low delay, circuit switched service are guaranteed cell transmission at a given rate. Packet mode users contend for the remaining cells with queuing and buffering determined by the traffic. Thus, the cell relay concept provides an inherent circuit-packet integration. One manufacturer's switching system** handles dedicated circuit switch service, connection-oriented (CO) packet switch service, and connectionless (CL) LAN service as shown in Figure 7-4.

7.1.3 Packetization

Packet switching has been used in the past primarily to achieve greater efficiency by time-sharing expensive transmission facilities used to transport bursty data traffic. The traditional packet switching technology is being supplanted with fast packet switching to take advantage of the large bandwidth and reliability of optical fibers. It is useful to compare the concepts of time division multiplexing (TDM), as used in digital circuit switching, and packet multiplexing as used in fast packet switching.

In TDM, the transport stream of bits is divided into frames of time slots and each slot in the frame is allocated to a particular user's data stream. Slot position in the frame identifies the user. In packet multiplexing, the bit stream is divided into packets and each packet labeled with a virtual channel identifier (VCI). The VCI identifies the packet allocated to a particular user. The ATM concept described in Section 3.4.4 is a special case of packet multiplexing where packets are all the same size, and are called "cells." The effective information transfer rate for a given user depends on the number of cells assigned to that user and not a recurring time slot. Therefore, the effective bit rate for a given user can be varied dynamically from zero to the full channel rate. An ATM network can, therefore, support a wide range of services as well as the traditional narrowband services in use today. It also conveniently handles bursty traffic by allocating bandwidth on demand.

**Siemens Stromberg-Carlsons' Electronic World Switching Digital (EWSC) System.

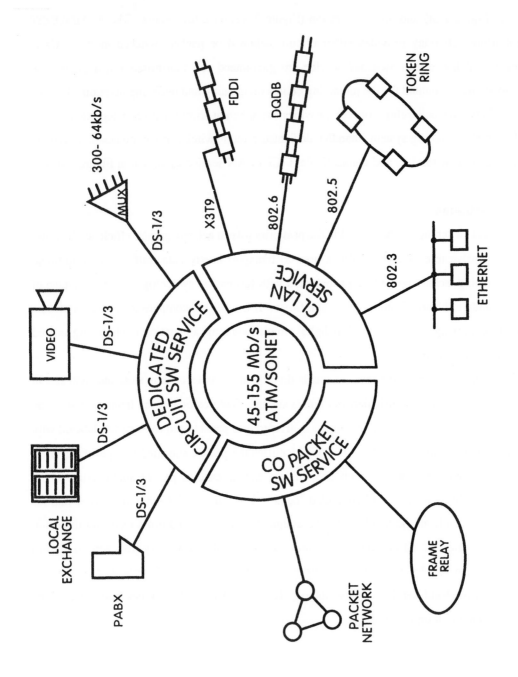

Figure 7-4. Services offered by one commercial switching system.

7.1.4 Privatization

Deregulation has led to the creation of private and virtual private networks for large businesses, and to competing providers of private network services. Virtual private networks appear to be private but are actually embedded (by software) in the public switched network.

There seems to be a number of reasons for the continuing proliferation of private networks. One reason is user requirements. In many cases, these requirements can only be met by developing a private network. Reliability, privacy, and security are three such requirements. Service quality, maintainability, customer control, and cost are others. There are several ways users can bypass the local exchange carrier, such as the use of satellite and microwave radio, in order to meet these requirements. A discussion of virtual private line networks (VPLN) and software defined networks (SDN) is given in Section 7.3.1.

According to Ryan (1991), there has been some shift back to the public network by larger enterprises for their communication and information processing because of the advanced capabilities of ISDN.

7.1.5 Standardization

The globalization of the economy has led to a need for international telecommunications standards. Many countries including the United States are extending their standards-making processes to the international arena. Figure 7-5 illustrates the evolution of a standard from a perceived need to the international standards area. Also see Section 4.2.

There are numerous standardizing efforts, both national and international, underway. The international efforts are key to the successful implementation of a truly global infrastructure. National variations to international standards, as well as different options and incompleteness of the standards themselves, still cause incompatibilities and make seamless networking across national boundaries difficult. Resolving these differences and incompatibilities continues to be a challenge to the standards community.

Figure 7-6 depicts how the Telecommunications Standards Committee T1 interacts with the U.S. CCITT National Committees to submit contributions for consideration by the International CCITT study groups and ultimately the Plenary Assembly. This shows the long and laborious process necessary to develop an international recommendation.

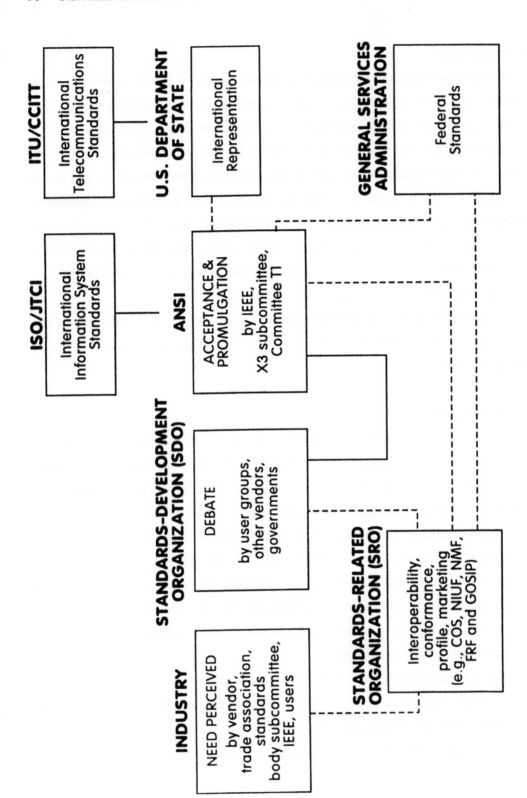

Figure 7-5. The evolution of a standard.

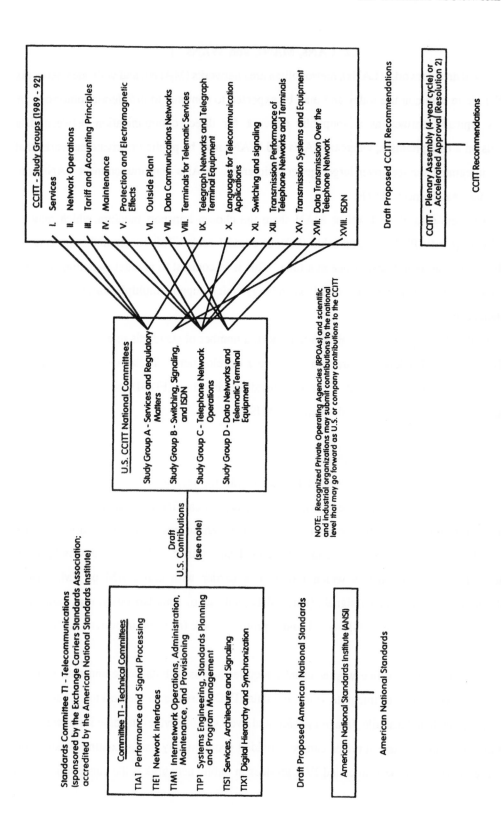

Figure 7-6. Development of ANSI standards and CCITT recommendations.

7.2 LANs, MANs, and WANs

Local area networks (LANs), metropolitan area networks (MANs), and wide area networks (WANs) are in common use today and may be expected to continue to proliferate tomorrow. A LAN is typically defined as a nonpublic network for direct communication between data terminals on a user premises, whereas MANs and WANs may be public or private networks that provide integrated services over larger geographic areas. MANs may interconnect LANs, and WANs may interconnect MANs.

All users of a given LAN or MAN share the chosen transport medium which may be twisted wire pair (WP), fiber optic cable (FO), or coaxial cable. This medium-sharing concept usually requires users to transmit one at a time. This is in contrast to parallel switching systems (e.g., asychronous transfer mode (ATM) switching) which support multiple users transmitting simultaneously.

Under the auspices of the IEEE 802 project, a number of standards relating to LANs and MANs have been developed or are currently under development. These are discussed in the following subsection along with the fiber distributed data interface (FDDI) standard promulgated by the American National Standards Institute (ANSI).

7.2.1 Local Area Networks (LANs)

Table 7-1 lists the IEEE Project 802 standards for various LANs. These LANs are distinguishable by the medium used (twisted wire pair, optical fiber, or coaxial cable), by the topology (bus, ring, star), by the medium access control (carrier sense, token passing), etc. Most LANs today operate at "baseband" and carry digital data traffic. There are, however, a few "broadband" LANs and MANs which modulate a carrier with either AM or FM. These broadband systems carry many signals (voice, data, and video) simultaneously, e.g., cable TV.

Figure 7-7 shows the OSI layer 1 and 2 protocol stack for some important LANs in use today.

Since the typical office worker usually requires access to both voice and data services, there is a growing trend to integrate voice and data (IVD) services to the desktop. These IVD services may include facsimile, image transfer, and even video services in some instances. This desktop integration of such services can be provided economically using existing twisted wire pair. The provision of these so called IVD services in public networks is the concern of CCITT

Table 7-1. IEEE Project 802 Standards

802.1	Describes network architecture concepts applicable to all networks including network management.
802.2	Describes connection-oriented and connectionless logical link control functions for layer 2.
802.3	Defines Ethernet protocol suite. Uses CSMA/CD MAC for use with a variety of physical medium dependent protocols.
802.4	Token bus MAC for use on 1 Mb/s coax and 20 Mb/s fiber. Originated by GM for MAP.
802.5	Token ring for use on TWP. Originated by IBM for use with a variety of physical medium dependent protocols for 1 Mb/s, 4 Mb/s, 16 Mb/s. Fiber version is FDDI.
802.6	Cell relay type LAN using 53 octets/cell for operating over dual bus using distributed queue for media access to the dual bus.
802.7	Broadband LAN - unapproved draft.
802.8	Fiber optic media.
802.9	Integrated voice and data (IVD) interface. This standard defines a unified access method that offers IVD services to the desktop from backbone networks. The operation is at 4 to 20 Mb/s over twisted wire pair using TDM frame of either 64 octets or 320 octets.
802.10	Network security standards for interoperable LANs.
802.11	Wireless LANs.

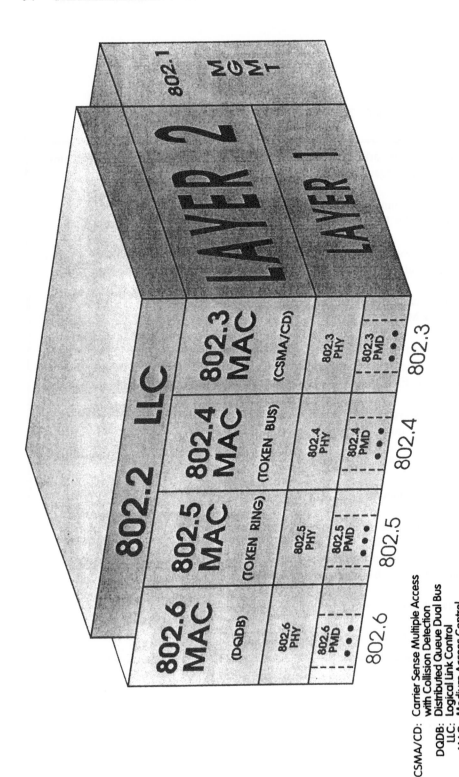

Figure 7-7. Protocol stack for LANs and MANs in use today.

CSMA/CD: Carrier Sense Multiple Access
          with Collision Detection
DQDB: Distributed Queue Dual Bus
LLC: Logical Link Control
MAC: Medium Access Control
MGMT: Management
PHY: Physical (medium independent)
PMD: Physical (medium dependent)

through the ISDN standards work. The LAN standards for IVD services are being developed by the IEEE 802.9 working group which is working to develop a standard for layer 1 and layer 2 with unshielded twisted wire pair as the transmission medium.

The IEEE 802.10 working group is concerned with defining a standard for interoperable LAN security including the secure data exchange protocol key management, and system security management.

The IEEE 802.11 group is developing a standard for wireless LANs. Wireless LANs operate without the traditional cabling techniques used today, but instead use radio links or AC power outlets to interconnect user terminals. Early systems operated in the upper UHF and SHF bands at approximately 900, 2,400, and 5,800 MHz using spread spectrum technology (SST). Power was limited to 1 watt which resulted in transmission distances of less than 1,000 feet.

At still higher frequencies (18-19 GHz), transmission speeds of 100 Mb/s are obtainable for digital termination service (DTS). Infrared wireless LANs are also feasible operating within line-of-sight.

Low-frequency carriers are used for radio systems that use the AC power facilities in a building. Maximum rates achievable are on the order of 100 kb/s.

In addition to these LANs, the ANSI Standard X3T9 defines a fiber distributed data interface (FDDI) for use as a LAN or MAN. This standard for optical fiber media defines physical and data link protocols for a dual counter-rotating token-ring type LAN operating at 100 Mb/s and capable of accommodating up to 500 nodes with a total distance of 100 km (about 62 miles) around the rings. A FDDI network is depicted in Figure 7-8 for a typical LAN application. Other example applications include interconnecting low-speed LANs that are dispersed in a campus-like environment, for interconnecting mainframe computers to mass storage devices and other peripheral devices, and for connecting integrated voice/data PBXs, computer hosts, and digitized video sources into a composite network for accessing ISDN.

The ANSI committee is also seeking to develop a single standard for 100 Mb/s FDDI signaling over both shielded twisted pair (STP) and unshielded twisted-pair (UTP) wiring (McLacklan, 1992).

Private automatic branch exchanges (PABXs) may also be used in lieu of a LAN. Located on a customer's premises, a PABX is basically a switch, star-connected to the customer

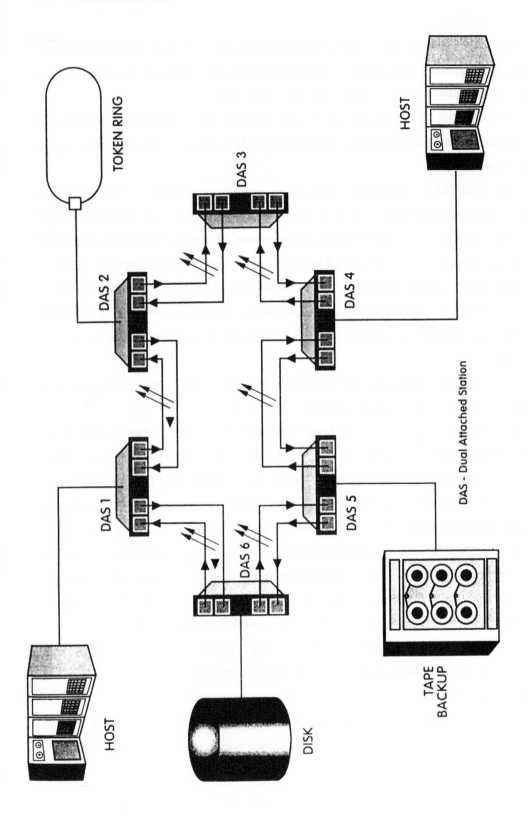

Figure 7-8. FDDI architecture with dual, counter-rotating ring.

terminals. Most PABXs today offer only telephone service but could support PC interconnections as well in the near future.

7.2.2 Metropolitan Area Networks (MANs)

MANs are used to interconnect LANs or to provide access to WANs. MANs are often commercial facilities that may be shared by several enterprises. MANs have the capability to allocate and deallocate vast amounts of bandwidths in extremely short time frames. They appear as a high-capacity line but users are charged only for the times when they are actually transmitting data. In the near term, technologies and services such as frame relay and IEEE 802.6-based SMDS offer the needed service, while broadband ISDN with ATM and SONET should provide the future service (Taylor, 1992). The IEEE 802.6 protocol suite for connecting LANs together over dual 45 Mb/s buses is available today. Bandwidth is 2 x 45 Mb/s now and expected to be 2 x 150 Mb/s in the future. Table 7-2 summarizes the important characteristics of major LANs and MANs in use today and in the foreseeable future.

7.2.3 Wide Area Networks (WANs)

WANs are the networks that interconnect LANs, MANs, or telecommunications terminals that are spread over large geographic areas, including global distributions. They are not defined by any single standard. Local exchange carriers (LECs) and interexchange carriers (IXCs) are considered WANs along with private corporate networks. WAN services range from dial-up analog services to ISDN. The switched multimegabit data service (SMDS) operating at 45 Mb/s is a switched data service available in the near term as a MAN and available in the future as a WAN. The synchronous optical network (SONET) operating up to 600 Mb/s and possibly higher is being standardized for future commercial service as a MAN or WAN. The SMDS concept is discussed in the following paragraphs. SONET was described in Section 3.4.5.

SMDS is a connectionless, cell-oriented, packet-switching service offered by a public network carrier. It is based on the distributed-queue-dual-bus (DQDB) concept of the IEEE 802.6 standard. Initially the SMDS will provide MAN-type service to interconnect LANs with high-speed (45 Mb/s) links. Eventually SMDS may provide wide-area network service for connections between local access transport areas (LATAs). The cell structure ensures compatibility with the future asynchronous transport mode (ATM) of the B-ISDN. Existing digital transmission

This page has been intentionally left blank.

Table 7-2. LANs and MANs Characteristics

Characteristic \ Identifier	Today					Tomorrow				
	802.3 CSMA/CD	802.4 Token Bus	802.5 Token Ring	SMDS	802.6 DQDB	ANSI X3T9 FDDI	SMDS Phase 3 (1995)	ATM/SONET	Wireless LANs	Wireless MAN
Architecture	Bus 10 Mb/s	Bus 1, 10, 20, Mb/s	Ring 1, 4, 16 Mb/s	Public High-Speed Data Service from LECs, 45 Mb/s (1992)	Looped Buses 45 Mb/s (90 Mb/s Capacity)	Dual Counter-Rotating Rings 100 Mb/s	DS-1, DS-3, National Service via LECs and IECs, 150 Mb/s	National/International Service and B-ISDN	Digital Cordless Data Terminals	Cellular Voice and Data
Access Control	CSMA/CD	Token Passing (Logical Ring)	Token Passing	SMDS Interface Protocol (SIP)	Distributed Queue Dual Bus (DQDB)	Token Ring MAC	SMDS Interface Protocol (SIP)	ATM Protocol (Link Layer)	CDMA or TDMA	CDMA and TDMA
Transmission Medium	• Coax • CATV Coax • Twisted Pair	• CATV Coax • Fiber	• Shielded Twisted Pair • Cable	• DS-1, DS-3 • SONET (future)	DS-3	Multimode Fiber	SONET (Fiber)	Fiber	Radio Frequency TBD	Radio (1.8-2.2 GHz, 400 and 800 MHz Bands)
Transmission Mode	• Baseband • Broadband • MAC Frames	• Broadband • MAC Frames ≤ 8140 Octets	Baseband	Baseband 53-Octet Cells (< 9188-Octet Packets)	Fixed-Length 53-Octets Cells	≤ 4500-Octet Frames	≤ 9188-Octet Packets of 53-Octet Cells	53-Octet Cells No Packetizing or Framing Required	Digital	Dual-Mode Digital and Analog Channels
Traffic Types	Computer Data	Computer Data or Manufacture Control	Computer Data	Images, Data, LAN Interconnect	Data with Hooks for Voice, Video	Data Only	Imagery, Data (LAN Interconnect)	Voice, Data, Video Multimedia	Data Only	Voice, Data
Service Types	Connectionless Logical Link Control (LLC)	Connectionless (LLC)	Connectionless (LLC)	Connectionless Network Service	Connectionless (LLC)	Connectionless (LLC)	Connectionless (LLC)	All (Connectionless, Connection-Oriented, and Circuit Switched)	?	Connection-Oriented (LLC)
Application	LAN Terminals	LAN for Manufacturing Automation	LAN	LAN/MAN Interconnect Service for Disaster Recovery	Private Networks on Customer Premises and as Alternative Access Provider or public-SMDS	High-Speed LAN for up to 500 Users	LAN/MAN Interconnect Service	B-ISDN Voice/Data/Video LAN/MAN Interconnect	Cordless PC Networking for LANs	Mobile Telephony and PCN
Comments	Commonly Known as Ethernet and Widely Used Today	GM Development Targeted for Industry/Manufacturing	IBM Concept Developed for Twisted Wire Pairs	Compatible with Existing LANs and Future B-ISDN Evolution	Well Matched to SONET	Campus Network to WANs	Based on 802.6 DQDB as Public Service Offering	Selected as International Standard for B-ISDN	New Technology; Low Penetration	6000 Cells and 6 Million Users in 1991

This page has been intentionally left blank.

facilities operating at DS-1 (1.5 Mb/s) and DS-3 (45 Mb/s) rates are currently used. Eventually 155 Mb/s and even rates up to 600 Mb/s could be used to carry SMDS via SONET. SMDS is usually considered to be a public MAN. This contrasts with FDDI which is intended primarily as an on-premises (e.g., a campus) LAN. The two concepts are actually complementary. FDDI LANs connected by SMDS may be used in the 1994 time frame (Weissberger, 1991a and b).

7.2.4 Interworking Devices: Repeaters, Bridges, Routers, and Gateways

LANs, MANs, and even WANs may be interconnected to extend geographic coverage using one of four basic devices: repeaters, bridges, routers, and gateways . The distinguishing characteristics between these devices are a function of the layer of the OSI at which they operate, as indicated in Figure 7-9. This seven-layer OSI model may be interconnected at layers 1, 2, and 3 using repeaters, bridges, or routers, respectively, or using gateways.

Repeaters operate at the physical level and simply regenerate signals transmitted across the network. They can interconnect LANs that use the same protocols. Bridges operating at level 2 connect networks such as LANs that use the same physical and link layer protocols. They generally have some intelligence for filtering and routing link layer frames to other network segments. Routers operating at level 3 have still more intelligence and can optimize packet routing to reduce congestion. Gateways permit the coexistence of OSI-based and proprietary products. The gateway connects different network architectures by performing a conversion at the application level. The gateway must utilize all of the layers of the proprietary architecture according to Stallings (1989).

7.3 Advanced Network Architectures

7.3.1 Virtual Private Line Networks (VPLN) and Software Defined Networks (SDN)

A VPLN is a private network provided on an as-needed basis to support a customer's application. The VPLN exists in the software embedded in the public switched network and may sometimes be called a SDN. One tariffed form of a SDN is offered by AT&T, and a software defined broadband network (SDBN) is planned for the near future (Wallace, 1992). For either, VPLN, SDN, or SDBN, the basic networking concept is time-sharing to make the public network appear private to the user. The advantages relative to leasing private lines are reduced cost and higher efficiency.

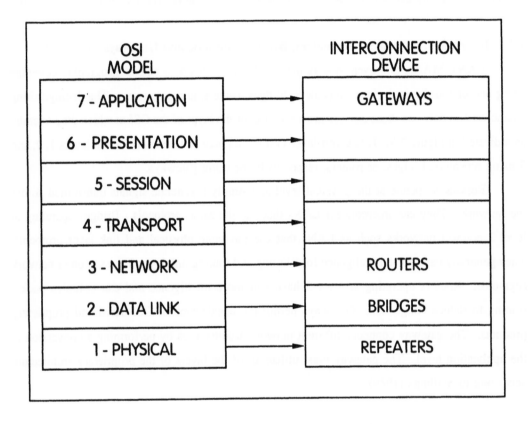

Figure 7-9. Mapping layers of the OSI model to corresponding interconnection devices.

Other transport systems that allocate and deallocate large amounts of bandwidth in extremely short times are expected to be needed in the future. This requirement has led to fast packet solutions for LANs, MANs, and WANs. A description of these bandwidth-on-demand transport systems was given in Section 7.2.

7.3.2 Intelligent Networks (IN) and Advanced Intelligent Networks (AIN)

The IN is a telecommunications network evolving from the public switched telephone network (PSTN) where service provisioning is provided by a service control architecture as depicted in Figure 7-10. The basic switching network contains a number of service switching points (SSPs) for switching user terminals. These SSPs are interconnected by Signaling System No. 7 (SS7). See Section 7.3.7 for a description of SS7. At call set-up, the SSP requests information about specific call handling from the network intelligence residing in service control points (SCP). The SCPs in turn are linked to a service management system (SMS) which is usually a commercial computer system with a number of remote peripherals. The SMS enables the network operator to manage and operate network services. The SMS may also be accessed by service providers to control, monitor, or modify service offerings.

Previously, when new services were offered by the switching system, it involved changes to thousands of different kinds of switches and took a long time. The IN allows new services to be introduced rapidly and efficiently through software changes to the centralized data bases and their associated operations support systems (OSS). The intelligence resides in on-line, real-time, centralized data bases, rather than in every switch, and is accessed through a packet-switched signaling network called SS7. Signaling networks based on SS7 provide the transport for IN services and call processing in local and backbone networks. The SS7 combines high performance, high reliability, and can respond rapidly to possible processor or link failures and congestion in the network. The infrastructure of IN is described by Robrock (1991) and by Claus et al. (1991).

The CCITT (1992) is developing draft recommendations for IN known as Capability Set 1 (CS-1). The CS-1 will permit the introduction of a wide range of advanced services with rapid service implementations and customization capabilities. Benchmark services being addressed by CS-1 are listed in Table 7-3. Features of CS-1 services which make up a service or represent a full service are listed in Table 7-4.

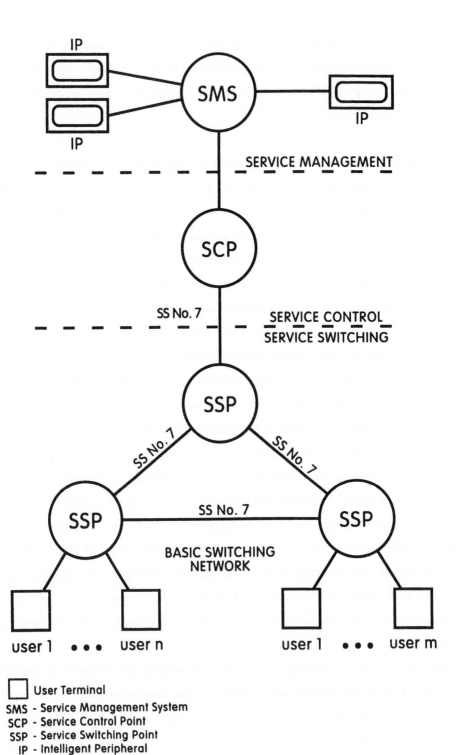

Figure 7-10. Intelligent network implementation.

Table 7-3. Benchmark CS-1 Services

Freephone (FPH)	Virtual Private Network (VPN)
User-Defined Routing (UDR)	Abbreviated Dialing (ABD)
Originating Call Screening (OCS)	Terminating Call Screening (TCS)
Call Forwarding (CF)	Call Distribution (CD)
Destination Call Routing (DCR)	Televoting (VOT)
Security Screening (SEC)	Premium Rate (PRM)
Split Charging (SPL)	Account Card Calling (ACC)
Credit Card Calling (CCC)	Automatic Alternative Billing (AAB)
Mass Calling (MAS)	Follow-Me-Diversion (FMD)
Conference Calling (CON)	Universal Access Number (UAN)
Malicious Call Identification (MCI)	
Completion of Call to Busy Subscriber (CCBS)	
Call Rerouting Distribution (CRD)	
Selective Call Forward on Busy/Don't Answer (SCF)	
Universal Personal Telecommunications (UPT)	

Table 7-4. Features of CS-1 Services

Reverse Charging	Call Distribution
Call Gapping	Call Limiter
Call Queuing	Call Screening (outgoing)
Call Screening (incoming)	Closed User Group
Customer Profile Management	Follow-Me Diversion
Origin-Dependent Routing	Customized Recorded Announcement
Time-Dependent Routing	User Prompter
Abbreviated Dialing	Authentication
Authorization Code	Off-Net Access
Off-Net Calling	Attendant
Mass Calling	Split Charging
Premium Charging	Private Numbering Plan
One Number	Customized Ringing
Call Logging	Personal Numbering
Call Forwarding	Multi-Way Calling
Call Waiting	Call Transfer
Meet-Me Conference	Consultation Calling
Call Hold with Announcement	
Automatic Call Back	

A paper by Duran and Visser (1992) describes the objectives of IN and includes an overview of CS-1. Work on future capability is expected to continue in order to include services that could occur during the active phase of a call, multimedia services, and for supporting topology management. A goal of the IN is to design standard interfaces which will facilitate the introduction of Open Network Architecture (ONA) into the public switched network in the United States. In the U.S., the pioneering work on IN was mostly done by Bell Communications Research (Bellcore) beginning in 1984. In 1989, Bellcore proposed an Advanced Intelligent Network (AIN) concept for the 1995 time scale (Bellcore, 1990). Implementing an IN or AIN concept is essential to future PCS and UPT systems discussed in Section 7.3.6.

7.3.3 Integrated Services Digital Network (ISDN)

The CCITT (1988a) defines ISDN as: "A network, in general evolving from a telephony integrated digital network, that provides end-to-end digital connectivity to support a wide range of services, including voice and non-voice services, to which users have access by a limited set of standard, multi-purpose user-network interfaces."

The ISDN recommendations define two interface rates for ISDN: a primary rate interface (PRI) and a basic rate interface (BRI). Both interfaces support simultaneous full-duplex voice and data using circuit-switched (voice) and packet-switched (data) connections on the same channel. Figure 7-11 indicates the ISDN interfaces.

The PRI is typically used to interconnect high-bandwidth devices such as mainframe computers, PBXs, and groups of lower bandwidth, basic rate lines with the central office digital exchange. The North American PRI is based on the DS1 transmission rate of 1.544 Mb/s. It consists of 23 64-kb/s B channels for voice and data, and one 64-kb/s D channel for signaling. Because a single D channel is used to handle all signaling, the other 23 channels are available for user data and voice transmission.

European PRI specifies 32 total channels (30B+1D+1 control) with an aggregate data rate of 2.048 Mb/s.

The BRI, which provides a composite bandwidth of 144 kb/s, is typically used to carry data to and from small end-user systems, such as voice/data workstations and terminal adapters for non-ISDN devices. It consists of two 64 kb/s information channels (B channels), which are used for voice and data, and one 16 kb/s packet-switched data channel (D channel), which can

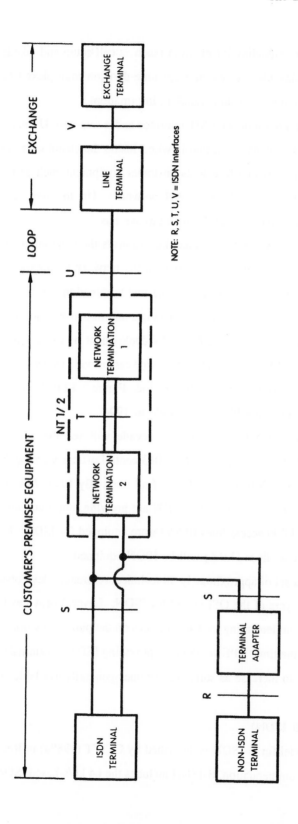

Figure 7-11. Recommendations for ISDN interfaces.

be used to carry either data or signaling information (such as call setup and takedown). The B and D channels are full-duplex bit streams, and are time-division multiplexed (2B+D) into a common stream that contains both user and signaling information.

The BRI consists of both a four-wire S/T interface and a two-wire U interface. The S/T interface standard is a four-wire connection that is wired inside the customer's home or office. Using the standard wall plug, this interface links customer equipment such as telephones, fax machines, and computer terminals with the ISDN network. Up to eight devices may be connected to this four-wire interface, which forms a passive bus.

The U-Interface (U) is a two-wire interface that connects the local telephone lines to the customer's home or office. The U interface is used in the network termination that links the customer premises with the local telephone line, in the line termination that links the telephone line with either a PRI or the central office digital exchange, and in two-wire terminal equipment.

Bellcore (1991) outlines the schedule for ISDN deployment by the seven RBOCs. This report estimates that by 1994, 61.9 million lines nationwide can provide ISDN service. Table 7-5 shows the implementation plan for the 1991 to 1994 period. Figure 7-12 indicates the availability of Signaling System No. 7 (SS7) and ISDN from 1990 to 1994.

According to the FCC, the first part of the decade will see the burgeoning of SS7 availability through BOC central offices. In 1990, 2,083 central offices were equipped for SS7, as were 36,706 access lines, which represent 34.7% of all lines. By 1994, 73% of BOC access lines will be equipped for SS7. The figures for ISDN are growing, but much more slowly. In 1990, 426 central offices and 496 access lines (0.5%) were equipped for ISDN. By 1994, 2,269 central offices and 2,218 access lines (1.9%) will be ISDN-equipped.

National ISDN-1 is a set of approximately 50 technical references developed by Bellcore that address the first three layers of the OSI model for ISDN. Layer 3 provides functional call control. The carriers began implementing ISDN-1 services on the public network for verification purposes in 1992. At the same time, CPE vendors are providing ISDN-1 terminals to customers. Implementation is expected to increase as software becomes generally available (Jones, 1991).

7.3.4 Broadband ISDN (B-ISDN)

The architecture model for B-ISDN is described by CCITT (1988a) in Recommendation I.327. According to this recommendation, B-ISDN includes the 64 kb/s-based ISDN capabilities,

Table 7-5. ISDN Implementation Plan

	1991		1992		1993		1994	
	LINES (MILLIONS)	% OF TOTAL LINES	LINES (MILLIONS)	% OF TOTAL LINES	LINES (MILLIONS)	% OF TOTAL LINES	LINES (MILLIONS)	% OF TOTAL LINES
Ameritech	2.2	15	3.36	22	8.0	51	11.15	70
Bell Atlantic	6.9	38	14.8	79	15.8	82	17.1	87
Bell South	3.1	17	5.6	30	8.0	41	10.5	52
NYNEX	1.3	8.3	3.88	24.5	5.27	32	5.47	33
Pacific Telesis	4.1	30	4.7	33	5.8	39	7.5	50
Southwestern Bell	1.6	12.9	2.0	15.6	2.1	15.9	2.2	16.3
US West	3.7	29	6.7	45	7.6	55	8.0	59

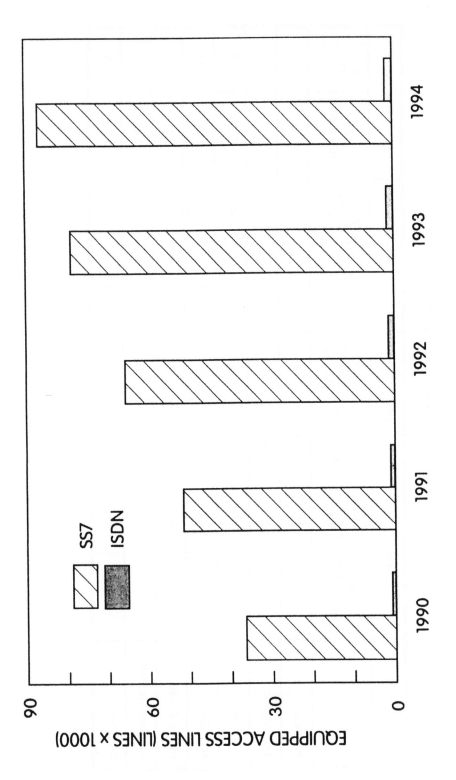

Figure 7-12. Availability of Signaling System 7 and ISDN central offices.

broadband capabilities, and control signaling from user to user, user to network, and between exchanges, as illustrated in Figure 7-13.

The B-ISDN is an ATM-based network. Transmission of B-ISDN signals in the access network is predominately over optical fiber using the synchronous digital hierarchy (SDH) defined by CCITT Recommendations G.707, 708, and 709 (CCITT, 1988e). These recommendations provide bit rates of approximately 155 Mb/s (STM-1), 622 Mb/s (STM-4), and 2.5 Gb/s (STM-16). Lower rates of 1.5 Mb/s and 45 Mb/s may be used initially to support ATM cell transport. The ATM multiplexor or switch adapts cell timing of incoming signals to the internal timing so that transmission links need not, in principal, be synchronized. However, initial implementations of ATM networks must support currently existing synchronous transport modes. The ATM/SONET concept was described in Section 3.4.

The B-ISDN trunk network consists of the following network elements: 1) virtual channel and virtual path switch (B-ISDN exchange), 2) ATM virtual path crossconnect, and 3) STM multiplexer crossconnect, as shown in Figure 7-14. A protocol reference model for B-ISDN is shown in Figure 7-15.

7.3.5 Internet and NREN

The Internet is a heterogeneous collection of computer networks organized in a hierarchy of networks connected through gateways, all using the transmission control protocol and internet protocol (TCP/IP), and all sharing common name and address spaces. Internet exists to facilitate the sharing of resources of participating organizations including government agencies, educational institutions, and private corporations. In January 1992, there were over 750,000 information processors on Internet linked by 5,000 networks with over 3 million users. The Internet evolved from ARPANET which was originally operated by DoD. The main backbone networks of Internet are MILNET and NSFNET which are mostly funded by Government grants, whereas the smaller networks are funded by other organizations. There are generally no per-user or per-message charges. Internet has grown almost exponentially since it first evolved in the late 1970's. An Internet Activities Board (IAB) is the general technical and policy oversight body. The IAB is taking steps to integrate Open Systems Interconnection (OSI) protocols into the Internet. These OSI protocols will coexist with TCP/IP and interoperability between OSI and TCP/IP is to be provided (Cerf and Mills, 1990).

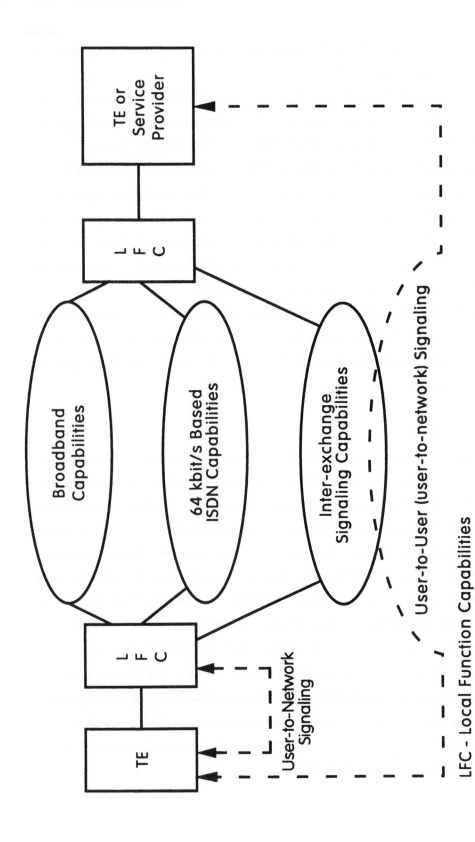

Figure 7-13.   Architecture model for B-ISDN.

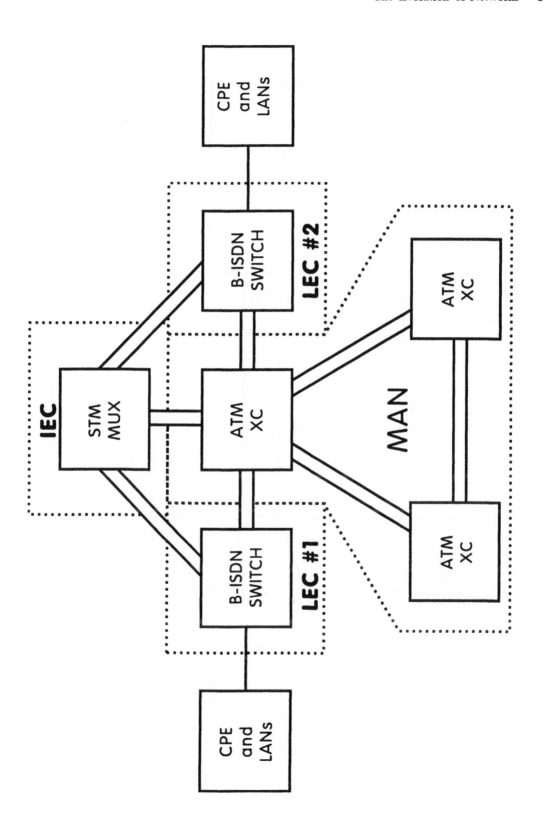

Figure 7-14. An example broadband ISDN (B-ISDN) trunk network.

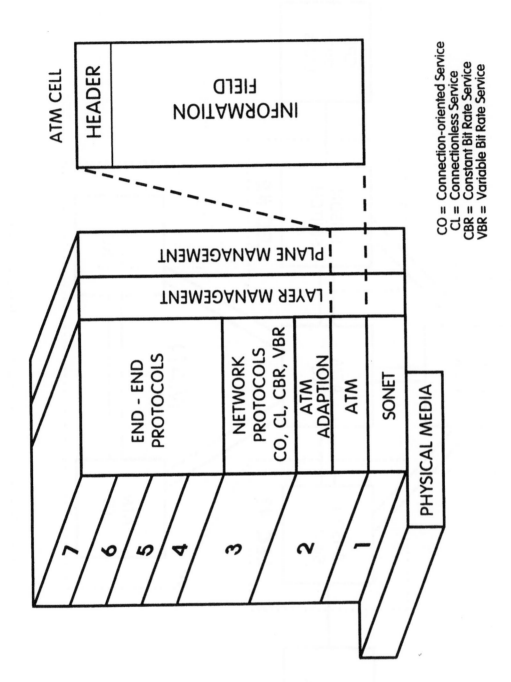

Figure 7-15. Protocol reference model for B-ISDN.

Martin (1991) lists the various resources, such as reference information, library documents, and other items available via Internet. In addition, the network provides services such as E-mail to a multitude of users. Finally, Clark et al. (1991) describes possible future directions for the Internet architecture and suggested steps towards the desired goals.

In the future, a multigigabit network known as the National Research and Education Network (NREN) may evolve from the existing Internet base. The description of NREN that follows was taken from Cerf (1990).

1.   There will continue to be special-purpose and mission-oriented networks sponsored by the U.S. Government which will need to link with, if not directly support, the NREN.

2.   The basic technical networking architecture of the NREN will include local area networks, metropolitan, regional, and wide area networks. Some nets will be organized to support transit traffic and others will be strictly parasitic.

3.   Looking towards the end of the decade, some of the networks may be mobile (digital, cellular).   A variety of technologies may be used, including, but not limited to, high speed Fiber Data Distributed Interface (FDDI) nets, Distributed Queue Dual Bus (DQDB) nets, Broadband Integrated Services Digital Networks (B-ISDN) utilizing Asynchronous Transfer Mode (ATM) switching fabrics as well as conventional Token Ring, Ethernet and other IEEE 802.X technology. Narrowband ISDN and X.25 packet switching technology network services are also likely play a role along with Switched Multi-megabit Data Service (SMDS) provided by telecommunications carriers. FTS-2000 might play in the system, at least in support of Government access to the NREN, and possibly in support of national agency network facilities.

4.   The protocol architecture of the system will continue to exhibit a layered structure although the layering may vary from the present-day Internet and planned Open Systems Interconnection structures in some respects.

5.   The system will include servers of varying kinds required to support the general operation of the system (for example, network management facilities, name servers of various types, e-mail, database and other kinds of information servers, multicast routers, cryptographic certificate servers) and collaboration support tools including video/teleconferencing systems and other "groupware" facilities.   Accounting and access control mechanisms will be required.

6. The system will support multiple protocols on an end-to-end basis. At the least, full TCP/IP and OSI protocol stacks will be supported. Dealing with Connectionless and Connection-Oriented Network Services in the OSI area is an open issue (transport service bridges and application-level gateways are two possibilities).

7. Provision must be made for experimental research in networking to support the continued technical evolution of the system. The NREN can no more be a static, rigid system than the Internet has been since its inception. Interconnection of experimental facilities with the operational NREN must be supported.

8. The architecture must accommodate the use of commercial services, private, and Government-sponsored networks in the NREN system.

A review of the evolution of Internet and NREN is given by Hart et al. (1992).

7.3.6 Personal Communications Network (PCN) and Universal Personal Telecommunications (UPT)

PCN is a new network expected to evolve from cellular mobile technology as an independent network to support hand-held personal voice and low-speed data communications terminals. It will interwork with the PSTN and ISDN networks. UPT is a service expected to evolve from many intelligent networks that support portable person-to-person telecommunications for voice and data, and including broadband services such as high-quality voice, data, facsimile, and video. UPT subscribers will have a personnel telecommunications identifier (PTI) and be able to receive calls at any terminal anywhere to which the subscriber has directed his calls. Thus, UPT would be available globally, with personal mobility provided and with a number associated with the person rather than a terminal.

The key to UPT is the intelligent network concept described in Section 7.3.2. UPT is concerned with the overall mobility of the human user and not so much with user equipment. Implementation of UPT implies a number of network capabilities including user identification, personal mobility, personalization, security, and confidentiality as described by Claus et al. (1991).

Personal communications networks (PCNs), also known as personal communications services or personal communications systems (PCS) are expected to have explosive growth by the mid 1990s. By the end of the decade, the users of PCN and other wireless services may exceed 25% of the U.S. population. A number of issues must be resolved before low-power, low-cost, digital PCNs can be deployed on a large scale. One issue discussed by Barnes (1991) involves spectrum allocation. Another related issue is the access technology to be used. The access choices include code division multiple access (CDMA), time-division multiple access (TDMA), and frequency-division multiple access (FDMA). See Viterbi (1991) and Schilling et al. (1991).

The ultimate future of PCN depends on field experiments and on the initial service acceptance. Tests are currently underway that involve many users, cells, and transmission environments.

The FCC has granted licenses in the 1850-1990 MHz frequency band for testing PCN systems using CDMA. The tests are intended to show that this band can be shared with microwave transmission users. Studies have shown that the demands for spectrum could be enormous. For example, one study by A. D. Little projected 60 million users 10 years after deployment (Mason, 1991). Others projected the numbers of users to be only half as large, i.e., 30 million. This compares with 11 million cellular subscribers today.

PCNs are dependent on the existing public networks such as the IN and AIN in the United States to tie together the wireless islands of coverage. AIN would provide mobility to users and expand network services to roaming customers.

Standards work in the U.S. for PCN is currently occurring in the ANSI T1P1 committee. What standards will emerge are yet undetermined.

Mobile satellite communications systems can also impact PCN (Lodge, 1991). Satellites can provide mobile services to large regions that cannot be served by terrestrial means (e.g., oceanic areas and sparsely populated regions). Experimental satellites and systems are in the advanced stages of deployment that could provide basic communication services, such as voice and low-rate data to very small terminals including hand-held units. Examples of these experimental satellites are ACTS (see Section 3.5), and the Personal Access Satellite System (PASS) proposed by the Jet Propulsion Laboratory (Sue, 1990).

7.3.7 Signaling and Network Management

Signaling systems provide remote control of network switches. The primary function is call control for voice and data transmission services. However, modern signaling systems also provide a number of advanced features and functions to the network user. Signaling is normally provided by a packet-switched network that is separated from the networks carrying voice and data. The CCITT-recommended signaling protocol that is central to intelligent networks (IN), cellular telephony, and ISDN architectures is known as Signaling System Number 7 (SS7). It provides a number of advanced services via the signaling network databases (CCITT, 1988b).

The overall objective of SS7 is to provide an internationally standardized, general-purpose, common-channel signaling system with five primary characteristics. First, it is optimized for use in digital telecommunication networks in conjunction with digital stored program control exchanges utilizing 64 kb/s digital channels. Second, it is designed to meet present and future information transfer requirements for call control, remote control management, and maintenance. Third, it provides a reliable means for the transfer of information in the correct sequence without loss or duplication. Fourth, SS7 is suitable for operation over analog channels and at speeds below 64 kb/s (e.g., 4,800 b/s). Finally, it is suitable for use on point-to-point terrestrial and satellite links.

SS7 is basically a packetized data network designed for transferring control information between processors in a telecommunications network, and is fast becoming the predominate method for controlling global networks (Jabbari, 1991). It is also a key element to intelligent networks, ISDN, and for UPT (see Sections 7.3.2, 7.3.3, and 7.3.6). The availability of SS7 in central offices from 1990 to 1994 was shown previously in Figure 7-12 of Section 7.3.3.

Network management systems are used to manage network resources in contrast to signaling systems that control the switches. They may both use the same network elements or be completely separated. The International Standards Organization (ISO) has defined five essential network management functions as follows:

- Configuration management which manages the state of the network

- Fault management which handles faults in the network

- Accounting management (billing) which charges for network resource usage

- Security management which manages security facilities

- Performance management which takes care of network performance.

The CCITT (1988d) Recommendation M.30 describes network management. As telecommunication networks become more complex the need for an effective management structure becomes important. The CCITT is developing a standard telecommunications management network (TMN) for this purpose in Recommendation M.30. The basic concept behind the TMN is to provide an organized structure to interconnect various types of operating systems and telecommunication equipment using a common architecture with standardized protocols and interfaces. The general relationship of a TMN to the telecommunications network is shown in Figure 7-16. Network management concepts, standards, and products are also described by Jennings et al. (1993).

7.4 Summary of Capabilities and Applications

Table 7-6 summarizes many of the networks discussed previously. Some pertinent characteristics of major networks in use today, and contemplated for tomorrow, are shown in this table. Critical trends in network architectures expected by the next decade along with some of the major issues to be resolved are discussed in Section 8.

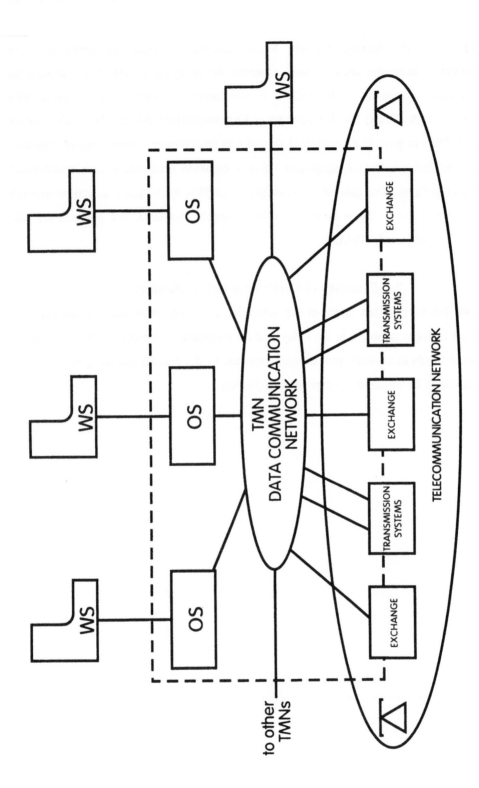

Figure 7-16. General relationship of a TMN to a telecommunication network.

Table 7-6. Major Network Evolution Summary

	Today					Tomorrow				
	PSTN	PDN X.25	ISDN	INTERNET	MOBILE RADIO	IN-1	SMDS	B-ISDN	NREN	PCN
Switching (packet, circuit, cell, or frame relay)	Analog and Digital (50% digital in 1991)	Packet Switch	Circuit, Packet	Packet Switch (Datagrams)		Circuit, Packet	Cell Relay	ATM Cell (Virtual) Circuit Switch at Link Layer	Not Yet Determined	
Transmission (fiber, copper, satellite, radio)	Copper, Fiber	Copper, Fiber Optic, Satellite	All (basic subscriber interface - 2B+D twisted pair)	All	Cellular Radio via Copper or Fiber Network	All Media	Fiber (45 Mb/s)	SONET/SDH Fiber (150 to 600 Mb/s)	Variety	Radio or Satellite with Fiber Interconnect
Signaling (in-band, SS7)	Analog #6	In-Band	SS7 20-70%	In-Band	In-Band	SS7	In-Band			
Penetration	135 Million Users in 1991	Unknown	RBOC Users 20-70% in 1994	3 Million Users Worldwide in 1990	6 Million Users in 1991	Unknown			Construction Beginning in 1993	
Nodes	~120 Nodes for AT&T	Unknown	Unknown	5000 Networks 570,000 Hosts	3 Million Users in 1992					
Subscriber Access and Data Rates	Data with Modems 300-9600 b/s	Dial-up: 9600 b/s Dedicated Lines: 56 kb/s to 1.5 Mb/s	$2B+D_1$ ($B = 64$ kb/s, $D_1 = 16$ kb/s) $2B+D_2$ ($B = 64$ kb/s, $D_2 = 64$ kb/s)	56 kb/s - 1.544 Mb/s Update to DS3, 45 Mb/s						
Network Layer Service [Connectionless (CL) or connection oriented (CO)]	CO	Primarily CO	CO/CL			CO	CL	CL and CO	Not Yet Determined	TDMA and CDMA
Service Offerings (voice, data, imagery, multimedia)	Primarily Voice; also Data via Modem	Data	Voice, Data	Data (file transfer, remote computing, E-mail)		Voice, Data with Special Features	Multimedia	Multimedia	Multimedia	Voice, Data to Mobile Users
Implementation Schedule	Currently Available 1993	Currently Available 1993	Currently Available 1993	Growth ~ 20% per Year		~ 1994	Available 1993	1995-2000	~ 1995	~ 1995

This page has been intentionally left blank.

8. The Critical Trends and Issues

"A scientist is a man who can find out anything, and nobody in the world has any way of proving whether he found it out or not, and the more things he can think of that nobody can find out about, why the bigger the scientist he is."

Will Rogers, 1930's

This section summarizes major trends and issues faced by those working in the information age early in the 21st century. Section 8.1 begins with a series of quotations from various authors working in the field of telecommunications. Then in Section 8.2, some projections are added with a list of seven critical areas expected to impact the telecommunications infrastructure ten years from now. In making these projections, one feels somewhat like the scientist defined above by Will Rogers. At least the aim is to follow Lenz's Law which states "When predicting the future it is better to be approximately correct than precisely wrong!" See Vanston et al. (1989). Some projections are obvious, others are nebulous due to fast-changing needs and technologies. Still others depend on innovative research yet to reach completion. Finally, Section 8.3 provides a list of some remaining issues that still need to be resolved as the 21st century is approached.

8.1 Pertinent Quotes

This report has addressed emerging and evolving telecommunications technologies, and has attempted, where it seemed appropriate, to project their trends and outcomes. This is important to those who must try to plan for the future. Questions remain, however, as to the accuracy of the predictions. The quotations below are intended to give the perspectives of some others who write books and articles on this subject, and who are called on by corporations and Government agencies to help sort out the elements of the telecommunications explosion. A number of these individuals have already been referenced, but the quotes given here serve as a nice summary for the salient ideas.

Taylor (1992), on desktop computing:

"The first key trend is the continuing shift in computing from host-centric to fully distributed networking. Equivalent to the shift from batch to interactive computing, this is only the second major evolution seen in this industry. This shift, fueled by the ready availability of high-performance low-cost workstations, is moving more and more of this intelligence in the network to the desktop. Indeed it is not uncommon for the desktop processor today to have as much raw computing power as a fairly large time-shared mainframe had just 10 years ago."

"Future networks will need to handle very bursty, high-volume, high-speed traffic with essentially no delay because interactions occur less frequently but require instantaneous response time for entire file transactions."

Tamarin (1988), on the role of the user:

"One very significant result of the new technologies have been the user's ability, given sufficient resources, to bypass the public network and build private network facilities. This is a trend that will surely continue, to the detriment of the telecommunications infrastructure, if traditional service providers do not become more responsive to user needs. Consequently, regulatory organizations need to be active in seeking user involvement in order to avoid user abandonment of the public network."

Mushin (1991), on privacy and security issues:

"The privacy protection and information security issue has been called the problem of the human gold fish. The telecommunications business, whose mission is to store, process, and carry information, has a primary responsibility with respect to this issue. The protection of information ownership and privacy against third parties is necessary whether the third party is government, major institutions, political groups, or criminals. In this area, the development of business and private cryptographic methods and equipment is essential."

Mushin (1991), on globalization of telecommunications:

"Telecommunications technologies link people all over the world. Helping poor people around the globe is a moral imperative for the telecommunications business.

Economically, telecommunications companies should stimulate economic growth in the less developed countries through direct and/or indirect investment, purchase of their goods, and active cooperation with national/international organizations.

Socioculturally, they should respect cultural traditions of other peoples.

They should also realize that they are required to protect foreign peoples as well as their own people from destructive transborder information flow threatening national defense or violating fundamentally different values and cultures."

Bonatti et al. (1989), on future network planning:

"Network planners must develop decision support systems needed for each of the stakeholding communications providers to assess the attractiveness of opportunities to adopt new telecommunications technology, and to commit plant expenditures, operations, and support systems necessary to accommodate new revenue-producing network services at forecasted levels of demands. As alternative services emerge, an increasing responsibility of the network planner will be to assist in decisions to withdraw from certain offerings and to abandon certain technologies in preference of newer opportunities. In summary, the dilemma of providing end-to-end service to increasingly sophisticated and demanding customers who are confronted by an array of options to serve their telecommunications needs, while recognizing that no individual provider can serve those needs without cooperative endeavors from partners, is likely to be the most important telecommunications challenge facing the industry in the 1990's."

Ryan (1991), on integrating circuit and packet switching:

"To successfully interface with all data communications equipment on the host or LAN side and to interconnect over ISDN with all other devices, next generation ISDN access equipment must provide comprehensive integrated protocol support.

Users increasingly demand integrated support of packet and circuit switching. The demand for integration is based on the trade-off between performance and cost. For an organization's low-bandwidth applications, packet-switching services are substantially more cost effective. For high-throughput, response-critical applications, the dedicated line of a circuit-switched connection is necessary."

LaBlanc (1992), on future services:

"In the coming decade, the importance of communication for the economic well-being of businesses, individuals, and the country as a whole will continue to increase. To meet the demands of the marketplace, extensive and innovative changes will occur. Today's vision of information superhighways and a national broadband network will become a reality.

Competition and technological innovation will increase the kinds of services available and the number of providers handling those services. Each of us may have our own personal communications number that will follow us to various locations, connecting us with our car phones, our offices, and our homes."

Brown (1985), on future services:

"Beyond ISDN, we can look forward to the day when customers, from the small family to the largest multi-national corporation, will have ready access to an array of information and communications services that once were dreamed of only by writers of science fiction.

At the same time, data, image, and video communications will eventually become as easy to use and as readily available as voice communication is today. For example, video or wideband channels available on demand for any purpose imaginable. Access to data bases around the globe -- perhaps accompanied by instantaneous translation of foreign languages into the customer's own language. An unlimited assortment of software-based services, including services involving networks that can be set up immediately to the customer's specifications."

Herr (1986), on universal information services:

"While ISDN provides a powerful start, customers are going to demand more: more integration, more control, and more bandwidth to carry an increasing volume of traffic that will be in image and video formats. People communicate with each other through multimedia formats -- wideband voice and image combinations using the larynx, ears, and eyes as audio and visual ports. Therefore, communications networks that support integrated voice and image combinations have a greater potential to achieve information productivity than integrated voice and data combinations alone. What's more, people need to be able to get information when they need it. It must be accessible rapidly. And since everyone has a range of communication needs that vary from time to time, the ability to provide services on demand becomes important to achieving information productivity.

...what we're envisioning is nothing less than turning communications networks into giant, distributed computers, with all of their power available to customers at their fingertips and under their control. We want to universalize the Information Age."

Claus et al. (1991), on IN and UPT:

"With the application of the intelligent network (IN) architecture, information technology is introduced in the network by way of specific service centers. These service centers concentrate the intelligence which is necessary for the support of the services, and they are linked with the switching network via powerful signaling links based on Signaling System No. 7. With this concentration of intelligence at a limited number of locations, services can be implemented quickly, can be cost-efficient, and can easily be modified to meet new market demands. ... one such new telecommunication service, the Universal Personnel Telecommunication (UPT) service will provide network independent personal identification allowing the relationship between terminal identity and user identity to be changed and thus providing a complete mobility across multiple networks."

Jacobsen (1992), on OSS and AIN:

"Today's operation support systems (OSS) are based on basic telephone service. Future OSSs must accommodate ISDN, SMDS, ATM, and the radical changes that will occur with AIN deployment. One of the goals of AIN is rapid service deployment, which is accomplished by placing general-purpose computers into the network that affect call processing and provide a switch-independent programming environment. The deployment of AIN will change the way telcos do business, which will significantly affect current OSS's that simply cannot support the integrated environment AIN needs. ...future OSSs must provide timely, accurate and cost-effective support for all telco activities. Future systems will use imaging, expert systems and even artificial intelligence."

Carpenter et al. (1992), on ATM/SONET standards:

"In the future, long distance digital transmission up to 2.4 Gb/s will be based on the Synchronous Digital Hierarchy (SDH) recommendations from the CCITT, including the American SONET standards. Additionally, Asynchronous Transfer Mode (ATM) is being defined internationally by the CCITT. ... ATM allows the definition of virtual paths through the network, with multiple virtual circuits within each virtual path. For non-data services, the virtual circuits will behave in the same manner as SDH circuits, and the underlying ATM 53-byte cell structure will be hidden in silicon by the ATM adaption layer. Various technical issues remain open at this time (March 1992) in the definition of the ATM adaption layer for data services."

Modarressi and Skoog (1990), concerning B-ISDN:

"In the second half of this decade, Broadband ISDN (BISDN) capabilities are likely to emerge in the network. BISDN provides a cell-based network infrastructure with extremely high-speed switching and transmission capabilities. Although it is not clear what the penetration rate of BISDN-based services will be in the telecommunication market place, it is quite clear that broadband technologies will become commercially available and new services using these capabilities will be offered. Given that the BISDN network will be based on asynchronous transfer mode (ATM) switching/transmission principles, and implemented on a ubiquitous optical fiber facility infrastructure, use of enormous bandwidth on demand will become not only technologically possible, but also economically feasible. Truly integrated multimedia services involving voice, high-speed data, image, and video will emerge and penetrate the business and residential markets with a potential to profoundly impact and transform the very fabric of those markets and the nature of the work place."

Cerf (1990), on the INTERNET and NREN:

"Looking towards the end of the decade some of the networks may be mobile (digital, cellular). A variety of technologies may be used, including but not limited to FDDI, DQDB, B-ISDN using ATM switching fabrics as well as conventional token ring, Ethernet, and other IEEE 802.X technology. Narrowband ISDN and X.25 packet switching technology network services are also likely to play a role along with SMDS provided by telecommunication carriers. FTS-2000 might play in the system at least in support of national agency network facilities.

The protocol architecture of the system will continue to exhibit a layered structure although the layering may vary from present day Internet and OSI structures in some respects."

Lyles and Swinehart (1992), on broadband networking:

"Traditionally, the speed and capacity of data communications networks are increased incrementally in response to incremental increases in network traffic. We see this trend today in the introduction of such services as ISDN, frame relay, and SMDS to support applications (e.g., electronic mail and file transfers) that constitute the bulk of the current traffic. ...recently, several bold and compelling activities have been launched to counter this evolutionary approach with a revolutionary one, designed to support important new applications whose voracious bandwidth requirements and delay sensitivities cannot be supported by today's networks."

Gant (1990), on SONET penetration:

"One of the backbone technologies to be rolled out in the 1990's is SONET. If it follows the course of most technology substitutions, its period of most significant growth will be from 1995-2000." See Figure 8-1.

Carpenter et al. (1992), on gigabit networking:

"Gigabits/s networking is on the verge of realization. The hardware technology is in the laboratories or the test beds. The software technology is appearing, and applications, especially multi-media networking and long-distance visualization, are taking shape. Even before the technology is in the market, standardization efforts, especially ATM, FDDI follow-on, HIPPI (High Performance Parallel Interface) and FCS (Fiber Channel Standard) are under way. Technical challenges remain but the biggest barriers, particularly outside the United States, will be political and economic. The research community should take the lead in breaking their barriers."

Kleinrock (1992), on propagation delay in gigabit networking:

"The major conclusion of this paper is to recognize that gigabit networks have forced us to deal with the propagation delay due to the finite speed of light. Fifteen milliseconds to cross the United States is an eternity when we are talking about gigabit links and microsecond transmission times. As we saw earlier, the propagation delay across the USA is forty times smaller than the time required to transmit a 1-Mb file into a T1 link. At a gigabit, the situation is completely reversed, and now the propagation delay is 15 times larger than the time to transmit into the link. We have moved into a new domain in which the considerations are completely reversed. We must rethink a number of issues. For example, the user must pay attention to his file sizes and how latency will affect

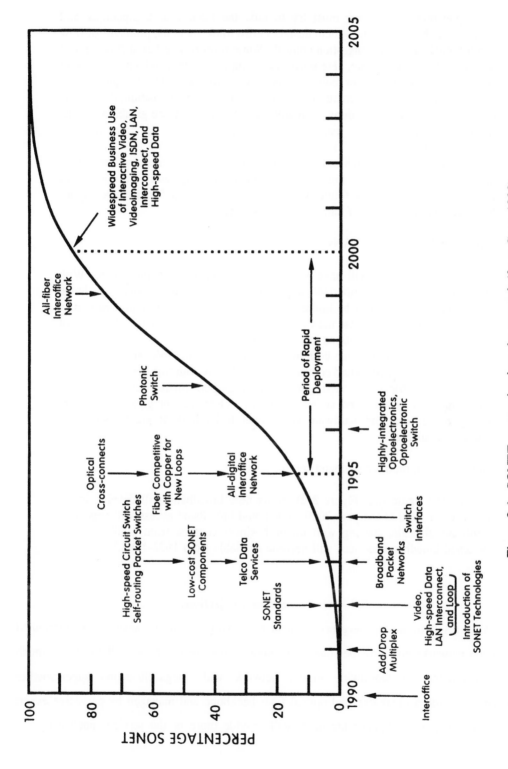

Figure 8-1. SONET penetration into the network (from Gant, 1990).

his applications. The user must try to hide the latency with pipelining and parallelism. Moreover, the system designer must think about the problems of flow control, buffering, and congestion control. Some form of rate-based flow control will help the designer here. He must also design algorithms which make rapid decisions if enormous buffer requirements are to be avoided. The designer cannot depend on global state information being available in a timely fashion; this affects his choice of control algorithms. In many ways, the user will see gigabit networks as being different from megabit networks; the same is true for the designer/implementer.

Much more research must be done before we can claim to have solved many of the problems that this new environment has exposed. We must solve these problems in the near future if we are to enjoy the benefits that fiber optics has given us in the form of enormous bandwidths."

Kung (1992), on gigabit LANs:

"Gigabit LANs will have a revolutionary impact on applications. With these networks many important applications (e.g., imaging and distributed computing) will no longer be limited by network speed. More importantly, new applications enabled by these networks will emerge, and will change the fields of computing and communications in a fundamental way.

Gigabit LANs provide both high-bandwidth (multi Gb/s) and low latency (tens of μs or less) end-to-end communication. LANs with this performance will pave the way for many next generation, high-performance computing and communication systems and new applications. Demands for gigabit LANs are widely recognized, and hardware component technology at gigabit speeds is basically available."

Timms (1989), on the broadband service schedule:

"The figure (our Figure 8-2) shows how broadband will develop in the period from 1980 to the year 2000. Integrated broadband access will appear on a substantial scale, starting with major business centers, from 1988 onwards. Switched broadband services will become available from 1992-1993."

8.2 Seven Ten Year Projections

As a result of the convergence of data processing and data transmission, distributed information networks now provide some new services, but are just now on the brink of much more. Ultimately, a worldwide digital network with megacell/second superswitches interconnected with a gigabit/second optical fiber backbone and managed by pervasive expert systems or artificial intelligence machines will provide more of the services needed by our

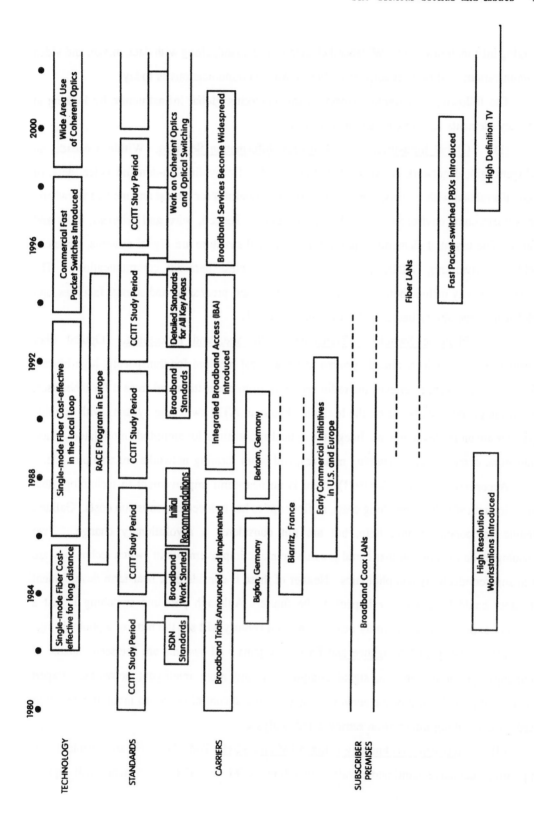

Figure 8-2. The development timetable for broadband (from Timms, 1989).

emerging information society. Wideband channels on-demand, along with data, image, and video communications will be as readily available as voice communications is today.

The following paragraphs expand on this emerging global infrastructure by looking at some key elements and making ten-year projections.

(1)   Global Infrastructure for Universal Information Services. Within a decade, an intelligent, global network is expected that is capable of providing on-demand voice, data, or video information with universal access (terrestrial or satellite) for almost anyone and anywhere. The infrastructure environment will be digital and packetized, wire and wireless, cells and cellular. The essential elements required for this global structure are digital systems, advanced intelligent networking (AIN), and some form of wireless personal communications. The infrastructure may include both public and private facilities, processing and storage facilities, and a global interoperability based on international standards.

(2)   Photonic Switching, Transmission, Storage, and Processing. Optical fiber transmission systems have already proven to be an ideal medium for transporting high-bit-rate information. As such, it relieves the radio spectrum from congestion and frees it for other uses. Optical integrated circuits are on the horizon. Photonic ATM switching systems with terabit/s capacities are under development. Integrated optical applications for amplitude modulators, phase modulators, demodulators, switching matrices, and beam forming networks are also possible.

According to Kleinrock (1991), there is no near-term improvement in technology that would dramatically drop switching costs, as fiber optics has reduced transmission costs. Gallium arsenide components would help, but are not considered a revolutionary change. Two technologies on the horizon that could provide dramatic improvements in switch technology are warm superconductivity and photonics. Neither of these is considered a near-term development but either could have a major impact in the future. Although photonic switching is just a laboratory experiment, it is entirely possible the 21st century will see the birth of a photonic age.

(3)   Artificial Intelligence and Expert Systems. AI systems are foreseen capable of continuous speech recognition, signal compression, and more intelligent networks. Expert systems will provide most of the network management and control functions involving decisions based on networking information retrieval and analysis.

(4)   Accessing Technology using ATM and SDH. Today's LANs are unsuitable for supporting interactive multimedia traffic; therefore, ATM or ATM-like switches will replace

LANs, and ATM technology will permeate future customer premises. This is because ATM is a parallel switching system that supports multiple users transmitting simultaneously. LAN users, in contrast, share the transport medium one at a time. Like today's PABXs, small ATM switches on the business premises, connected by high-speed trunks to public backbone networks, will serve all of the business needs. Frame relay, SMDS, and FDDI will be fading from the scene.

The synchronous digital hierarchy standard for high-speed transmission will dominate the global infrastructure in ten years.

(5)   Wireless Systems. Wireless systems include mobile systems such as cellular systems, PCNs, cordless phones, paging, and wireless LANs. Media include terrestrial radio and satellite. In the next ten years, the industry expects many new wireless services to emerge as more and more people are on the move. Wireless LANs are just appearing now. Wide-area wireless private networks can be implemented with central transceivers and mobile equipments. Standards are currently being developed for the next-generation public cellular systems. This includes a possible all-digital cellular system based on spread spectrum technology (Krechner, 1991).

Cellular, satellite, and microwave systems to expected to continue to evolve, but each into its own specialty area of application.

For personal communications, a PCN will evolve and compact, fully-featured terminals will provide complete access to services anytime anywhere.

(6)   Multimedia Workstations and Communications. The organization, storage, manipulation, and distribution of information are all functions critical to the success of most businesses. As long as the importance of the content, not the conduit, is recognized, these workstations will have beneficial applications, including interactive education, multimedia conferencing, medical, and a host of others.

Already, PCs offering advanced capabilities such as color, audio, full-motion color, and video are available. It is estimated that by 1995 approximately 200 million of these PCs will have been sold.

(7)   Software Evolution for Parallel Programming. In ten years, heterogeneous networks of workstations will be common. Programmers will develop programs of different types and languages to interact in this environment. To do this, new programming methods will evolve using formalized procedures to develop, proof, and test each program.

8.3 Issues and Actions Impacting Telecommunications

This section includes four major issues: 1) standardization, 2) Government policies, 3) the telecommunications infrastructure, and 4) privacy, security, and fraud. A further breakdown for each follows.

8.3.1 Standardization

Standardization issues include those concerning the user's role in the entire standards-setting process, which has been largely indirect. Solving technical interface problems is hard. In a national multivendor environment, it's more difficult. And, solving hard technical problems in the international arena is almost impossible due to the large number of participants and issues involved. National interests may have to take a back seat and world markets taken into account when selecting what standards should be adopted.

8.3.2 Government Policies

The list of policy issues includes

- Issues involving security and emergency actions critical to the national welfare

- Policies of Department of State concerning international standards

- Regulatory policies of the FCC (e.g., ONA)

- Judicial policies concerning interexchange carriers and RBOCs (e.g., COMP III)

- Spectrum management policies

- Congressional policies affecting users (e.g., Communications Act of 1934).

8.3.3 Telecommunications Infrastructure

Issues in this category include

Bypass Issues. The user has a number of options (e.g., satellite, microwave radio, PCN) which provide capabilities to bypass the public network with a private network either in the local area or totally. There are a number of reasons for bypass. One is cost, another service and performance. Sharing the public network may be detrimental due to the different traffic types (e.g., continuous, intermittent, or interactive). Bypassing the RBOCs provides competition in the local area, but at the same time may be detrimental to the public telecommunications infrastructure. This could become a policy issue.

Survivability and Reliability Issues. These issues include the complex problems of cross-connecting competing networks to make a whole, and then providing adequate network management for this whole. Resolving survivability and reliability issues is of paramount importance to the user whose entire business depends on the infrastructure.

8.3.4 Privacy, Security, and Fraud

The issues in this category include

- Intellectual property protection

- Equitable access

- Customer control

- Computer virus protection

- Call screening and related features

- Caller identification and related features

- Network management

- Security technologies

- Authentication

- Access control

- Data confidentiality

- Data integrity

- Non-repudiation

- Vulnerability of software-controlled networks and systems.

8.3.5 A Comment Concerning Issues

Its noteworthy how often the user appears in every category. User concerns and needs are always an issue. When service providers, equipment suppliers, policy makers, and standards developers are unresponsive to users' needs, users tend to go elsewhere.

9. Conclusions and Recommendations

"Would you tell me, please, which
way I ought to go from here?" asked Alice.
"That depends a good deal on where you
want to get to," said the Cat.
"I don't much care where," said Alice.
"Then it doesn't matter much which way
you go," said the Cat.

Lewis Carroll, Alice's Adventures in Wonderland

Telecommunications has gone from POTS to PCS, and from PCs to integrated workstations and multimedia desktop terminals. Switching systems have advanced from analog to digital, from switching packets and cells to possibly photons. LAN technology is moving from coax to fiber, and even to wireless. Traffic, once carried by circuit switched or dedicated lines, now travels in packets and cells. Information rates have gone from kilobits/s and megabits/s to gigabits/s and even terabits/s.

Computers have more capacity, operate faster, are cheaper and smaller. Networks are faster, cheaper, more reliable, and service oriented. Terminals are more integrated, more intelligent, more usable, more mobile, and there are more of them.

New technologies were introduced with new names like ATM, SDH, SMDS, AI, VSAT, FDDI, and HDTV. Network technologies like 802.6 LANs, DBDQ MANs, and AIN WANs were explored. New standards from organizations like CCITT, ISO, ANSI, T1, and IEEE, based on models called OSI, ONA, and TMN, are leading to new layered architectures called B-ISDN, PCN, and UPT using new standard protocols. These networks range in scope from local to regional and from national to international. Some are public, some are private, some are regulated, some are not.

Today's cable delivers 100,000 times as much bandwidth to users as their telephone, yet costs about the same (Rosner, 1990). LAN speeds have increased from 100 Mb/s to 1 Gb/s, memory capacity from 1 Mbit/chip to 500 Mbit/chip, yet costs constantly decrease. Transmission speeds over fiber are going from Gbit/s to Tbit/s and processing speeds from MIPS to BIPS, and still costs continue to fall, foreshadowing spectacular achievements in the future.

The report shows how PCN, VSATS, and microwave radio might bypass LECs to provide competition in that last holdout. How fiber and ATM and SONET could provide a host of new broadband services. How PCN could add mobility so that the global infrastructure could provide services to anyone, anywhere, anytime. It also indicates why the generation, organization, manipulation, and distribution of information is so critical to our nation's future.

It is predicted that by the year 2002 there will be a global infrastructure, managed and controlled by artificially intelligent machines, providing universal service using photonic switching and storage systems, and connecting distributed computing systems or multimedia terminals operating at gigbit rates. Others agree! Rappaport (1991) believes that a wireless revolution is forthcoming. Over the past few years, interest in wireless has been spectacular. Cellular radio had 50% growth rates, pagers 70%, and PCN research has been intense. Still others like Robinson (1992) note that "the intelligent global communication network is the engine that will provide simplified and universal access to any medium for anyone, anywhere ... on demand." Kleinrock (1992) concurs and states that "telecommunications, based on some of the most exciting technologies available, is changing rapidly and influencing almost every aspect of business, commerce, education, health, government, and entertainment."

Like Alice, one wonders which way to go from here. There are a number of trends yet to be explored and much more detail to be added for the most promising ones. It is recommended that future efforts be used to expand on the level of detail. The basic technonogies that should be examined for their "impact" are listed below.

- Fiber optics and photonics

- Synchronous optical network (SONET)

- Asynchronous transfer mode (ATM) - WANS & LANS

- Broadband-ISDN

- Personal communications systems (PCS)

- Switched multi-megabit data service (SMDS)

- Wireless LANs

These technologies will form the foundation for most of the telecommunication services which will be prevalent ten years hence.

In the past, our nation's economy depended on natural resources like oil, minerals, and crops. In the future, the strategic resources will not come just from the ground but will be the ideas and information that come out of our minds. The new technologies that support telecommunications provide the means to use and expand this new strategic information resource efficiently and on a global basis.

10. Addendum
Telecommunications Trends Update (1994)

The report on Present Status and Future Trends in Telecommunications was prepared in the latter part of 1992 and early 1993. The evolution of telecommunications depends on many factors such as user demand, technologies available and their costs, government regulation, security/privacy issues, and last but not least the existing plant both inside and outside structures. However, as noted in Section 1.2 of this report concerning the future infrastructure of telecommunications 'the constantly changing structure may be expected to continue in almost every aspect of the network.' In this addendum, we highlight a few of the major changes that have occurred in the past year and present an updated view of where the future telecommunications infrastructure is headed.

Wireless personal communications (WPC) is emerging as the next technical revolution that could transform the way society works and plays. WPC takes on many forms (satellite-, microcell-, and picocell-based) as it progresses along a number of paths (technical, standards, regulatory, and political). At the same time, there are certain restrictions to the advance of WPC due to spectrum constraints and performance limitations as noted in the previous section.

There are two basic questions that need to be addressed.

(1) How great an impact will be introduction of advanced wireless technologies such as PCS, wireless LANs, and wireless PBXs have on the future telecommunications infrastructure?

(2) How can the security and privacy issues associated with wireless transmission be resolved in view of the openness of the media?

To answer the first question it is necessary to take a broad view of the future trends in communications and the infrastructures that tie it all together. This infrastructure is constantly undergoing change as a result of new technologies, deregulation, economic factors and users needs. Our society is increasingly dependent on information - how it is created, processed, and transferred. Thus, it is important to explore the emerging telecommunications infrastructure that plays such as important role in our nation's future.

The future infrastructure is based on our perception of the overall evolution of networks and services as discussed below.

In previous work on future trends in telecommunications, a complex diagram of networking services versus time was developed. This diagram depicted the convergence of LANs, MANs, and WANs into a single merged entity capable of carrying the traffic for a multiplicity of functions and services. A simplified version appears as follows:

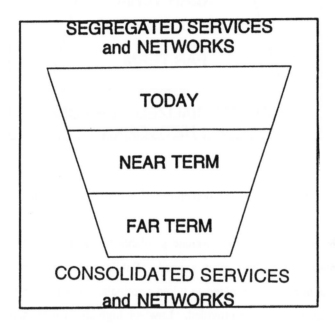

Figure 1. Providers perception of network consolidation.

Thus, in the far term there is one vast seamless telecommunications infrastructure providing almost every service and function every user needs. Voice, data, and video would all be integrated to flow over this one vast consolidated network.

Reflecting on this diagram one colleague suggested it was upside down and that the opposite might occur. In his view many more new networks would evolve each providing its own unique service with optimum performance to very specialized groups of users. Each distinct network would stand alone and in only a few cases would they be connected together. This concept would appear as follows:

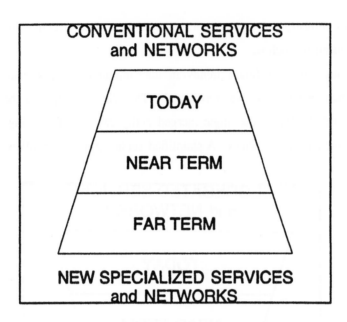

Figure 2. User's perception of network segregation.

Here more and more services become available as the need arises by adding special networks each optimized to best meet a need.   Public and private networks continue to proliferate, wireless terminals are added to wired terminals, and simultaneous combinations of voice, data, or video services are provided. Low -or high-speed data, slow- and fast-motion video, as well as a host of multimedia services will all be provided some on separate optimized networks, while others combine on a single network. Wireless access traffic in both cases would likely be only a small percentage of the traffic carried by the networks. Figure 3 depicts the complexity of one possible evolving architecture.

The most likely scenario probably lies somewhere between the two concepts outlined above.   Instead of a single network providing many services or a proliferation of several specialized networks, there will be a few major networks (like the Public Switched Telephone Network, Public Data Networks, Internet, and ultimately the National Information Infrastructure) handling the majority of voice and data traffic with other specialized networks (public and private) evolving separately each with its own unique capabilities.

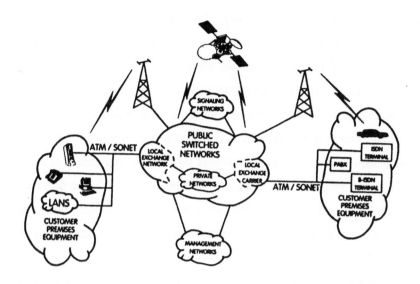

Figure 3. Complexity of the evolving telecommunications infrastructure.

One good example of an existing data network is Internet. The Internet is heterogeneous collection of computer networks organized in a hierarchy of networks connected through gateways, all using transmission control protocol and internet protocol (TCP/IP) and all sharing common name and address spaces. Internet today has several million subscribers with almost 1 million processors on the network.

By the end of the decade, a multimegabit network known as the National Research and Education Network (NREN) may evolve from Internet. By late 1990, some of the NREN access may be mobile. A variety of technologies may be used including, but not limited to, high speed Fiber Data Distributed Interface (FDDI) nets, Distributed Queue Dual Bus (DQDB) nets, Broadband Integrated Services Digital Networks (B-ISDN) utilizing Asynchronous Transfer Mode (ATM) switching fabrics as well as conventional Token Ring, Ethernet and other IEEE 802.X technology. Narrowband ISDN and X.25 packet switching technology network services are also likely play a role along with Switched Multi-megabit Data Service (SMDS) provided by telecommunications carriers. FTS-2000 might play a role in the structure, at least in support of Government access to the NREN, and possibly in support of national agency network facilities.

Another possible infrastructure may be similar to the broadband structure being standardized for B-ISDN (see Section 7.3.4). The current administration is also supporting a similar concept known as the National Information Infrastructure (NII). It has been said that "everyone is behind NII but no one knows what it is." It is obviously more than just a superhighway for data. NII is expected to evolve by a "practical merging of computer, communications, software, and information industries. The resulting infrastructure will be a seamless, dynamic web of networks, information appliances, information resources, and people that will simultaneously carry limitless amounts of information in a variety of formats including voice, video, test, and multimedia to unlimited locations"[*] The evolution of wireless structures such as mobile/portable satellite communications, worldwide untethered communications, and wireless offices along with personal communication service all may be building blocks leading toward NII.

Most of the interconnecting links between LAN, MAN, and WAN nodes in future networks such as NII would use ATM carried by SONET over optical fiber. See Sections 3.4.4 and 3.4.5. Almost every service imaginable would be provided, wireless and wired, with the flexibility to add new services as desired. The structure would be based on national and international standards, operate at several hundred megabits per second with one switching fabric in both circuit and packet switching modes. Data, voice, and video would be carried on the same network using bandwidth on demand.

The development of any infrastructure including NII and B-ISDN requires contributions from both the private and the government sector. The private sector role will undoubtably predominate but the government must still insure that information resources are available to all at affordable prices. Government action may also be necessary to ensure information security, reliability, and restorability requirements are addressed.

Below we present several important trends that could have major impact on the future telecommunications infrastructure. These projections have evolved from various sources including conversations with the staff at the Institute for Telecommunication Sciences in many of the references to this report.

---

[*]Kay, K.R., The NII: More than just a data superhighway, Telecommunications, January 1994, page 47.

- Future networks must accommodate existing applications and services as well as new ones that may not even be envisioned or conceived in the future. They must be capable of integrating fiber, coax, copper, and wireless transport facilities.

- Users will have more control over network resources, bandwidth allocations, and network management as core intelligence migrates to the network periphery. Excess bandwidth available in fiber transmission systems will be used for these signalling, control, and management purposes.

- Synchronous streams of small packets transmitted at rates ranging from 50 MHz to 2.5 Gb/s over synchronous optical networks (SONET) will be shared by many diverse users (voice, data imagery, etc.) using an asynchronous transfer mode (ATM). Future ATM/SONET networks will employ fiber transmission with photonic amplification, switching, and processing.

- Wireless systems will evolve to serve users anytime and anywhere who do not stay in fixed locations. This will cause a major change in the philosophy for the use of wired and wireless facilities. Broadband traffic such as televisioned destined for fixed locations will move to wires and cables whereas narrowband voice or data destined for mobile locations will be one radio. Cellular radio will evolve toward ubiquitous microcells for ubiquitous personal communications capable of reaching mobile individuals using their personal number wherever they are.

- Applications requiring transmission rates of gigabits (billions-of-bits) per second or more to individual workstations will proliferate. Optical LANs with gigabit/s capacity will evolve and ultimately be interconnected via broadband ISDN facilities.

- Future terminal functions will increase to include handwriting, image, or speech recognition and synthesis for interface control. Multimedia technologies will permit users to combine imagery with data and speech for a host of new applications including education, marketing, and entertainment. Many different multifunctional terminals with numerous software packages will access ISDN or B-ISDN for a variety of information services. Pocket-sized smart terminals and wireless PCS technology will provide users with voice, data, and images anywhere and at anytime.

Considering the total infrastructure it appears that wireless systems may carry only a small percent of the total traffic. However, they still pose some particularly challenging problems concerning spectrum usage, privacy, and security that must be addressed.

Table 1 illustrates several key evolutionary trends in telecommunication transmission/switching technology, networking, and wireless technologies and for customer premises equipment (CPE) with related services.

During the past year, other significant developments have occurred in the telecommunications arena. This includes a number of mergers, both real and contemplated, including AT&T and McCaw, Bell Atlantic and TCI, Southwestern Bell and Hanser Properties, U.S. West and Time Warner, Bell South and Prince to name just a few. In addition, court decisions have impacted the industry. For example, the decision invalidating the ban on telco provision of video programming on First Amendment grounds. The FCC licensing of PCS spectrum by auction and cable re-regulation may also have major impact. Even more changes are occurring even as this is written. One can hardly envision what more the future holds. As Alice would say, "Where ought we go from here?"

Table 1. Present and Future Trends in Telecommunications

	Switching/Transmission	Networking	Wireless Technologies	Services and CPE
1970s	First Fiber Optic Transmission T-Carrier Shared Media LANs Electronic Switching with Stored Program Control	Analog POTS In-Band Signaling #6 802.3 and 802.4 LANs Common Channel Signaling	Inmarsat-A (1976) Launched for Mobile TC Cellular Concept Proposed (1971) AMPS Intro (1998)	Microelectronics Telephone, TV Time Share Computers
1980s	Computer Controlled Switching Common Channel Signaling Digital Cross Connects Digital Switching Packet Switching & Transmission Optical Fiber Transmission	ISDN Standards (1988) AT&T Divestiture (1984) PSTN Converting to Digital Public Data Network with X.25 Data Modems 300 b/s - 2400 b/s Internet has 3M Users Worldwide DQDB LANs FTS-2000	Tone Alert Paging Land Mobile Radio Cellular Telephones Digital Microwave Alpha/Numeric Pagers GSM Standard Introduced Enhanced Page/Messaging Cordless Phones	Facsimile in Wide Use PC Sales Exceed 10 Million Microprocessors HDTV Standards Compact Discs Video Rentals Electronic Mailbox Personal FAX
Early 1990s	First Fiber Transmission @ 3.4 Gb/s Frame Relay Transmission Fast Packet Switching Asynchronous Transfer Mode Synchronous Optical Network	FDDI (ANSI X3T9) DNHR with 120 Nodes on PSTN RBOC ISDN Penetration 20-70% SMDS (Phase 2) IN-1 Initial Intro Intelligent Network	Digital Cellular Cordless Phones Wireless LANs Air-to-Ground Telephones Mobile Radio 6M Users ACTS Launched PCS Frequency Auction (FCC) PDAs Introduced	Microprocessor @ 10 MIPS LANs & MANs Operating @ 100 Mb/s Multimedia Terminal Available Mobile Satellite Services Interactive Voice/Data Multifunction Terminals
Late 1990s	Local SS#7 LAN/MAN Fast Packet Fast Packet Self Routing	Advanced Intelligent Network National Research and Ed. Network Universal Personal Telecommunication National Information Infrastructure	Wireless LANs/PBX Widespread MSAT Services FPLMTS Microcell/Picocell Portables LEO Satellites Launched	ISDN Videophones Video Mail Services HDTV Systems Proliferate Video Teleconferencing Integrated Voice/Data Workstations Hypermedia Standards
2000s	Wireless Subscriber Loop ATM/SONET Proliferation Optical Amplifiers	ATM LANs Gigabit Networking Dynamic Bandwidth Control B-ISDN Introduced Photonic Transport Fiber Access to the Home	Wireless Offices Global PCS DDAs Proliferate	Multimedia Terminal Proliferate Paperless Offices Broadband (B-ISDN) Workstations Interactive CAD/CAM Automatic Speech Recognition
2010s	Photonic Switching and Transmission	Global Network via B-ISDN B-ISDN Widely Available	Multimedia PCS Voice and Data Services via PCNs	Optical Processing Interactive Video ED & Distribution Services with User Control Virtual Reality

11. References

Amarel, S. (1991), AI research in DARPA's strategic computing initiators, *IEEE Expert*, June, pp. 7-11.

Anonymous (1990), Fiberworld overview, *Telesis*, Vol. 17, Nos. 1 and 2.

Anonymous (1991), Under a blue sky, Editorial, *Communications Week*, April.

Anunasso, F., and I. Bennion (1990), Optical technologies for signal processing in satellite repeaters, *IEEE Communications Magazine*, February, pp. 55-64.

Asatani, K., K.R. Harrison, and R. Ballart (1990), CCITT standardization of network node interface of synchronous digital hierarchy, *IEEE Communications Magazine*, August, pp. 15-20.

Ballart, R., and Y. Ching (1989), SONET: now it's the standard optical network, *IEEE Communications Magazine*, March, pp. 8-15.

Barnes, G. (1991), PCN versus microwave: the spectrum battle, *Telecommunications*, December, pp. 38 and 42.

Bellcore (1990), Advanced intelligent network, Release 1, Baseline Architecture, Special Report SR-NPL-00155, March, Livingston, NJ.

Bellcore (1991), ISDN deployment data, Special Report SP-NWT-002102, October, Livingston, NJ.

Bonatti, M., S.S. Katz, C.F. Newman, and Varvaloucas (1989), Guest editorial, *Journal on Selected Areas of Communications*, Vol. 7, No. 8, October, pp. 1161-1165.

Bourne, J., and J. Roth (1985), Technology as a network enabler, *Telesis*, No. 3, pp. 5-11.

Brown, C.L. (1985), Reach for a new goal: universal information services, *Telephone Engineer and Management*, November, pp. 69-71.

Brule, J., and I. Ebert (1990), Fiberworld: an overview, *Telesis*, Vol. 1/2, pp. 5-17.

Cargill, C.F. (1989), *Information Technology Standardization*, (Digital Press, Bedford, MA), Chapters 10-13.

Carpenter, B.E., L.L. Landweber, and R. Tirler (1992), Where are we with gigabits, *IEEE Network*, March, pp. 10-13.

Catlett, C.E. (1992), In search of gigabit applications, *IEEE Communications Magazine*, April, pp. 42-51.

CCITT (1988a), Recommendations of the IXth Plenary Assembly, Blue Books on ISDN, Fascicles III.7, III.8, and III.9, Geneva, Switzerland.

CCITT (1988b), Recommendations of the IXth Plenary Assembly, Blue Books on Signaling System Number 7, Fascicles VI.7, VI.8, and VI.9, Geneva, Switzerland.

CCITT (1988c), Recommendations of the IXth Plenary Assembly, Blue Books on Open System Interconnection (OSI), Fascicle VIII.4, Geneva, Switzerland.

CCITT (1988d), Recommendations of the IXth Plenary Assembly, Blue Books on Network Management, Fascicle IV.1, Geneva, Switzerland.

CCITT (1988e), Recommendations of the IXth Plenary Assembly, Blue Books on General Aspects of Transmission Systems, Fascicle III.4, Geneva, Switzerland.

CCITT (1992), Principals of intelligent network architecture, Draft Recommendations I.312-I.329.

Cerf, V. (1990), Thoughts on the national research and education network, Network Working Group, Request for Comments No. 1167, July.

Cerf, V., and K. Mills (1990), Explaining the role of GOSIP, Network Working Group, Request for Comments No. 1169, August.

Chandy, K.M., and C. Kesselman (1991), Parallel programming in 2001, *IEEE Software Magazine*, November, pp. 11-20.

Clark, D., L. Chajun, V. Cerf, R. Braden, R. Hobby (1991), Towards the future internet architecture, Network Working Group, Request for Comments No. 1287, December.

Claus, J., F. Lucas, and G.K. Helder (1991), New network architectural functionality, *Telecommunications Journal*, Vol. 58-XII, pp. 907-914.

Communications Act (1934), U.S. Code, Title 47, Para. 151.

COS (1987), COS protocol support, Version I, Corporation for Open Systems, COS/SFOR 87/0004.01, McLean, VA.

Daddis, G.E., and H.C. Lorng (1989), A taxonomy for broadband integrated switching architecture, *IEEE Communications Magazine*, Vol. 27, No. 5, May, pp. 32-42.

Duran, J.M., and J. Visser (1992), International standards for intelligent networks, *IEEE Communications Magazine*, February, pp. 34-42.

Ekas, W. (1991), Photo-nic finish, *TE&M Magazine*, December, pp. 39-43.

Elgen, D.J. (1990), Narrowband and broadband ISDN CPE directions, *IEEE Communications Magazine*, April, p. 39.

Fahey, M. (1991), Study sees $4.5 billion loop market in a decade, *Lightwave Magazine*, October, pp. 6-7.

FCC (1970), Computer I Decision, 28 FCC 2d at 291.

FCC (1973), Computer I Decision, 40 FCC 2d at 293.

FCC (1977), Computer II Decision, 77 FCC 2d at 384.

FCC (1986), Computer III Decision, Mimeo numbers 86-252 and 86-253, June.

Federal Standard 1037B (1991), Glossary of Telecommunication Terms, General Services Administration, Office of Information Resources Management, Washington, DC.

Fluent, M. (1987), Pursuing the dream: Mountain Bell's Phoenix ISDN trial, *Connections*, March, p. 6.

Fraser, A.G. (1991), Design considerations for a public data network, *IEEE Communications Magazine*, October, pp. 31-35.

Gant, S. (1990), White paper to management, *Network Management*, May, p. 44.

Gold, E.M. (1992), Video conferencing closes in on the desktop, *Network Management*, January, pp. 42-46.

Gould, J.D., S.J. Boires, and C. Lewis (1991), Making usable useful productivity, *Communications of the ACM*, Vol. 34, No. 1, January, p. 77.

Green, H.H. (1983), United States vs AT&T, Modified Final Judgment 552F. Supp 131, 229.

Grubb, J.L. (1991), The travelers dream come true, *IEEE Communications Magazine*, November, pp. 48-51.

Hart, J.A. R.K. Reed, and F. Bar (1992), The building of the internet, *Telecommunications Policy*, November.

Herr, T.J. (1986), The impact of technology divestiture and deregulation, International Wire and Cable Symposium, Reno, NV, November.

IEEE (1992), Special issue on gigabit networks, *Communications Magazine*, Vol. 30, No. 4, April, 124 pp.

Irven, J.H., M.E. Nilson, T.H. Judd, J.F. Patterson, and Y. Shibata (1988), Multimedia information services: a laboratory study, *IEEE Communications Magazine*, June, pp. 27-44.

Jabbari, B. (1991), Common channel signaling system No. 7 for ISDN and intelligent networks, *Proc. IEEE*, Vol. 79, No. 2, February, pp. 155-169.

Jacobsen, G.M. (1992), An inevitable evolution, *Telephony*, January, pp. 20-28.

Jansky, D.M., and M.C. Jeruchim (1987), *Communication Satellites in the Geostationary Orbit* (Artech House, Inc., Norwood, MA).

Jaske, R.P. (1991), Using VSATs for energy industry applications, *Telecommunications*, December, pp. 27-90.

Jennings, R.D., R.F. Linfield, and M.D. Meister (1993), Network management: a review of concepts, standards, and current status, NTIA Report 93-295, April.

Jones, S. (1991), National ISDN-1, *Telecommunications*, September, pp. 18-25.

Katz, J.E. (1990), Social aspects of telecommunications security, *IEEE Technology and Society Magazine*, June/July, pp. 16-24.

Kleinrock, L. (1985), Distributed systems, *Communications of the ACM*, Vol. 28, No. 11, November, pp. 1200-1213.

Kleinrock, L. (1991), ISDN - The path to broadband networks, *Proc. IEEE*, Vol. 79, No. 2, February, pp. 112-117.

Kleinrock, L. (1992), The latency/bandwidth tradeoff in gigabit networks, *IEEE Communications Magazine*, Vol. 30, No. 4, April, pp. 36-40.

Korpi, N.A. (1991a), Evaluating frame relay - Part 1, *Telecommunications*, October, pp. 54-58.

Korpi, N.A. (1991b), Evaluating frame relay - Part 2, *Telecommunications*, November, pp. 41-46.

Krechner, K. (1991), A review of U.S. mobile communication standards, *Telecommunications*, July, pp. 43-44.

Kung, H.T. (1992), Gigabit local area networks: a systems perspective, *IEEE Communications Magazine*, Vol. 30, No. 4, April, pp. 79-89.

LaBlanc, R.E., (1992), Special Report: when competitors become allies, *TE&M Magazine*, January, p. 6.

Linfield, R.F. (1990), Telecommunications networks: services, architectures, and implementations, NTIA Report 90-210, December.

Lodge, J.H. (1991), Mobile satellite communications systems: toward global personal communications, *IEEE Communications Magazine*, November, pp. 24-30.

Lyles, J.B., and D.C. Swinehart (1992), The emerging gigabit environment and the role of local ATM, *IEEE Communications Magazine*, April, pp. 52-58.

Madrid, J., S. Sheldon, and G. Cheadle (1991), Not just another digital hybrid, *TE&M Magazine*, August, pp. 42-45.

Martin, J. (1991), There's gold in them thar networks! or searching for treasure in all the wrong places, Network Working Group, Request for Comments No. 1290, December.

Mason, C. (1991), Spectrum tops all PCS debate, *Telephony*, December, pp. 9-10.

Matheson, R.J. (in press), Spectrum conservation: adjusting to an age of plenty, NTIA, Boulder, CO.

Mayo, J.S. (1985), The evolution to universal information services, *Telephony*, March, pp. 40-46.

Mayo, J.S., and W.B. Marx, Jr. (1989), Introduction to universal information services, *AT&T Technical Journal*, March/April, pp. 2-4.

McLacklan, G. (1992), 100 Mb/s copper vote nears, *LAN Computing Magazine*, Vols. 3/4, April, pp. 1.

McQuillan, J.M. (1991), Cell relay switching, *Data Communications Magazine*, September, pp. 58-69.

Medwinter, J.E., and M.G. Taylor (1991), The reality of digital computing?, *IEEE Lightwave Communication Systems Magazine*, May, pp. 40-44.

Millar, C. (1991), All light now, *IEEE Review*, January, pp. 35-91.

Miska, P.A., M. Peshavaria, G.M. Reed, and R.J. Wilson (1992), The emerging world of wireless communications, *AT&T Technology*, Vol. 6, No. 4.

Mayers, W. (1991), The drive to the year 2000, *IEEE Micro*, February, pp. 10-74.

Modarressi, A.R., and R.A. Skoog (1990), Signaling system No. 7: a tutorial, *IEEE Communications Magazine*, July, p. 34.

Mushin, L. (1991), Social responsibilities of the telecommunications business, *IEEE Technology and Society Magazine*, Summer, pp. 29-30.

Noam, E. (1988), The future of the public network: from star to the matrix, *Telecommunications*, March, pp. 58-65.

National Telecommunications and Information Administration (1991), Telecommunications in the age of information, NTIA Infrastructure Report, Special Publication 91-26, U.S. Department of Commerce.

Pelton, J.N. (1988), ISDN: satellites versus cable, *Telecommunications*, June, pp. 35-80.

Rappaport, T.S. (1991), The wireless revolution, *IEEE Communications Magazine*, November, pp. 52-65.

Robinson, C.A. (1992), Digital optical science bears smart pocket communicator, *Signal Magazine*, June, pp. 28-36.

Robrock, R.B. (1991), The intelligent network - changing the face of telecommunications, *Proc. IEEE*, Vol. 79, No. 1, January, pp. 7-20.

Rosner, R. (1990), Structure wideband communications: looking at the future, *Journal of Network Management*, Spring, pp. 16-21.

Ryan, R.J. (1991), Back to the future, *Telephony*, January, pp. 30-32.

Schilling, D.L., L.B. Milstern, R.L. Pickholtz, M. Mullback, and F. Miller (1991), Spread spectrum for commercial applications, *IEEE Communications Magazine*, April, pp. 66-79.

Shumate, P.W. (1989), Optical fibers reach into homes, *IEEE Spectrum*, February, pp. 43-47.

Smith, R. (1992), Cautious-not complacent, *TE&M Magazine*, January 15, pp. 31-39.

Stallings, W. (1989), Internetworking: a guide for the perplexed, *Telecommunications*, September, pp. 25-30.

Stallings, W. (1990), CCITT standards foreshadow broadband ISDN, *Telecommunications*, March, pp. 29-41.

Stallings, W. (1992), The role of SONET in the development of broadband ISDN, *Telecommunications*, April, pp. 21-24.

Sue, M.K. (1990), A satellite-based personal communications system for the 21st century, Proc. of the Second International Mobile Satellite Conference, Ottawa, Canada, June.

Tamarin, C. (1988), Telecommunications technology applications and standards, *Telecommunications Policy*, December, pp. 323-331.

Taylor, S.A. (1989), Plain talk about frame relay, *Network Management*, January, pp. 55-60.

Timms, T. (1989), Broadband communications: the commercial impact, *IEEE Networks Magazine*, July, pp. 10-15.

Vanston, L.K., R.C. Ling, and R.S. Wolf (1989), How fast is new technology coming? *Telephony*, September, pp. 49-52.

Vickers, R., and T. Vilmansen (1987), The evolution of telecommunications technology, *IEEE Communications Magazine*, July, pp. 6-18.

Viterbi, A. (1991), Wireless digital communication: a view based on three lessons learned, *IEEE Communications Magazine*, September pp. 33-36.

Wallace, R. (1992), AT&T lays plans for broadband SDN, *Network World*, March 23, pp. 1 and 44.

Warr, M. (1991), The pursuit of photons, *Telephony*, December 23, p. 22.

Weissberger, A.J. (1991a), Comparing alternative fast-packet network technologies, Part I, *Telecommunications*, September, pp. 58-65.

Weissberger, A.J. (1991b), Comparing alternative fast-packet network technologies, Part II, *Telecommunications*, October, pp. 53-69.

Wright, D.L., J.B. Balombin, and P.Y. Shon (1990), Advanced communication technology satellite (ACTS) and potential system applications, *Proc. IEEE*, Vol. 78, No. 7, July, pp. 1165-1175.

Appendix
Possible Broadband Services in ISDN

ANNEX A to Recommendation I.121 (CCITT, 1988a)

Service classes	Type of information	Examples of broadband services	Applications	Some possible attribute values [g) h)]
Conversational services	Moving pictures (video) and sound	Broadband [b) c)] video-telephony	Communication for the transfer of voice (sound), moving pictures, and video scanned still images and documents between two locations (person-to-person) [c)] — Tele-education — Tele-shopping — Tele-advertising	— Demand/reserved/permanent — Point-to-point/multipoint — Bidirectional symmetric/bidirectional asymmetric — (Value for information transfer rate is under study)
		Broadband [b) c)] videoconference	Multipoint communication for the transfer of voice (sound), moving pictures, and video scanned still images and documents between two or more locations (personne-to-group, group-to-group [c)] — Tele-education — Tele-shopping — Tele-advertising	— Demand/reserved/permanent — Point-to-point/multipoint — Bidirectional symmetric/bidirectional asymmetric
		Video-surveillance	— Building security — Traffic monitoring	— Demand/reserved/permanent — Point-to-point/multipoint — Bidirectional symmetric/unidirectional
		Video/audio information transmission service	— TV signal transfer — Video/audio dialogue — Contribution of information	— Demand/reserved/permanent — Point-to-point/multipoint — Bidirectional symmetric/bidirectional asymmetric
	Sound	Multiple sound-programme signals	— Multilingual commentary channels — Multiple programme transfers	— Demand/reserved/permanent — Point-to-point/multipoint — Bidirectional symmetric/bidirectional asymmetric

	Data	High speed unrestricted digital information transmission service	— High speed data transfer
			  — LAN (local area network) interconnection
			  — Computer-computer interconnection
			— Transfer of video and other information types
			— Still image transfer
			— Multi-site interactive CAD/CAM
				— Demand/reserved/permanent
				— Point-to-point/multipoint
				— Bidirectional symmetric/bidirectional asymmetric

		High volume file transfer service	— Data file transfer	— Demand — Point-to-point/multipoint — Bidirectional symmetric/bidirectional asymmetric

Service classes	Type of information	Examples of broadband services	Applications	Some possible attribute values [g), h)]
Conversational services (continued)	Data (continued)	High speed teleaction	− Realtime control − Telemetry − Alarms	
	Document	High speed Telefax	User-to-user transfer of text, images, drawings, etc.	− Demand − Point-to-point/multipoint − Bidirectional symmetric/bi-directional asymmetric
		High resolution image communication service	− Professional images − Medical images − Remote games and game networks	
		Document communication service	User-to-user transfer of mixed documents [d)]	− Demand − Point-to-point/multipoint − Bidirectional symmetric/bi-directional asymmetric
Messaging services	Moving pictures (video) and sound	Video mail service	Electronic mailbox service for the transfer of moving pictures and accompanying sound	− Demand − Point-to-point/multipoint − Bidirectional symmetric/uni directional (for further study)
	Document	Document mail service	Electronic mailbox service for mixed documents [d)]	− Demand − Point-to-point/multipoint − Bidirectional symmetric/uni directional (for further study)
Retrieval services	Text, data, graphics, sound, still images, moving pictures	Broadband videotex	− Videotex including moving pictures − Remote education and train-ing − Telesoftware − Tele-shopping − Tele-advertising − News retrieval	− Demand − Point-to-point − Bidirectional asymmetric
		Video retrieval service	− Entertainment purposes − Remote education and train-ing	− Demand/reserved − Point-to-point/multipoint [f)] − Bidirectional asymmetric
		High resolution image retrieval service	− Entertainment purposes − Remote education and train-ing − Professional image commu-nications − Medical image communica-tions	− Demand/reserved − Point-to-point/multipoint [f)] − Bidirectional asymmetric
		Document retrieval service	"Mixed documents" retrieval from information centres, archives, etc. [d), e)]	− Demand − Point-to-point/multipoint [f)] − Bidirectional asymmetric
		Data retrieval service	Telesoftware	

Service classes	Type of information	Examples of broadband services	Applications	Some possible attribute values [g], [h]
Distribution services without user individual presentation control	Video	Existing quality TV distribution service (PAL, SECAM, NTSC)	TV programme distribution	– Demand (selection)/permanent – Broadcast – Bidirectional asymmetric/unidirectional
		Extended quality TV distribution service – Enchanced definition TV distribution service – High quality TV	TV programme distribution	– Demand (selection)/permanent – Broadcast – Bidirectional asymmetric/unidirectional
		High definition TV distribution service	TV programme distribution	– Demand (selection)/permanent – Broadcast – Bidirectional asymmetric/unidirectional
		Pay-TV (pay-per-view, pay-per-channel)	TV programme distribution	– Demand (selection)/permanent – Broadcast/multipoint – Bidirectional asymmetric/unidirectional
	Text, graphics, still images	Document distribution service	– Electronic newspaper – Electronic publishing	– Demand (selection/permanent – Broadcast/multipoint [f] – Bidirectional asymmetric/unidirectional
	Data	High speed unrestricted digital information distribution service	– Distribution of unrestricted data	– Permanent – Broadcast – Unidirectional
	Moving pictures and sound	Video information distribution service	– Distribution of video/audio signals	– Permanent – Broadcast – Unidirectional
Distribution services with user individual presentation control	Text, graphics, sound, still images	Full channel broadcast videography	– Remote education and training – Tele-advertising – News retrieval – Telesoftware	– Permanent – Broadcast – Unidirectional

Notes to Table

a) In this table only those broadband services are considered which may require higher transfer capacity than that of the H_1 capacity. Services for sound retrieval, main sound applications and visual services with reduced or highly reduced resolutions are not listed.

b) This terminology indicates that a re-definition regarding existing terms has taken place. The new terms may or may not exist for a transition period.

c) The realization of the different applications may require the definition of different quality classes.

d) "Mixed document" means that a document may contain text, graphic, still and moving picture information as well as voice annotation.

e) Special high layer functions are necessary if post-processing after retrieval is required.

f) Further study is required to indicate whether the point-to-multipoint connection represents in this case a main application.

g) At present, the packet mode is dedicated to non-realtime applications. Depending on the final definition of the packet transfer mode, further applications may appear. The application of this attribute value requires further study.

h) For the moment this column merely highlights some possible attribute values to give a general indication of the characteristics of these services. The full specification of these services will require a listing of all values which will be defined for broadband services in Recommendations of the I.200-Series.

REFERENCE

CCITT (1988a), Recommendations of the IXth Plenary Assembly, Integrated Services Digital Network ISDN, Vol. III, Fascicle III.7, Melbourne, Australia, November, pp. 47-51.

Acronyms and Abbreviations

AC	Alternating Current
ACSE	Association Control Service Element
ACTS	Advanced Communications Technology Satellite
ADPCM	Adaptive Pulse Code Modulation
AI	Artificial Intelligence
AIN	Advanced Intelligent Network
AM	Amplitude Modulation
AMSC	American Mobile Satellite Corporation
ANSI	American National Standards Institute
ARPA	Advanced Research Project Agency
AT&T	American Telegraph and Telephone Company
ATM	Asynchronous Transfer Mode
AWO	Asian Workshop for OSI
b/s	Bits per Second
B-ISDN	Broadband Integrated Services Digital Network
Bellcore	Bell Communications Research
BER	Bit Error Rate
BIPS	Billion Instructions per Second
BOC	Bell Operating Company
BRI	Basic Rate Interface
BSA	Basic Serving Arrangement
BSE	Basic Service Element
C	Codec
CATV	Cable Television
CBR	Constant Bit Rate
CCIR	International Radio Consultative Committee
CCITT	International Telegraph and Telephone Consultative Committee
CCS	Common Channel Signaling
CD	Compact Disk
CDMA	Code Division Multiple Access
CEI	Comparably Efficient Interconnection
CL	Connectionless
CLNS	Connectionless Network Service
CNS	Complementary Network Services
CO	Connection-Oriented
COS	Corporation for Open Systems
CPE	Customers Premises Equipment
CPU	Central Processing Unit
CS	Capability Set

CSMA/CD	Carrier Sense Multiple Access with Collision Detection
CT	Cordless Telephone
DAS	Dual Attached Station
DDD	Direct Distance Dialing
DoD	Department of Defense
DQDB	Distributed Queue Dual Bus
DS1	Digital Signal at 1.544 Mb/s
DS3	Digital Signal at 44.76 Mb/s
DTE	Data Terminal Equipment
DTS	Digital Termination Service
EDI	Electronic Data Interchange
EIA	Electronic Industries Association
ESP	Enhanced Service Provider
ESS	Electronic Switching System
EWOS	European Workshop for OSI Standardization
FAX	Facsimile
FCC	Federal Communications Commission
FCS	Fiber Channel Standard
FDDI	Fiber Digital Data Interface
FDMA	Frequency Division Multiple Access
FIPS	Federal Information Processing Standard
FM	Frequency Modulation
FO	Fiber Optic
FTAM	File Transfer, Access, and Management
FTS	Federal Telecommunication System
FTTC	Fiber to the Curb
FTTH	Fiber to the Home
Gb/s	Gigabits per Second (10^9 b/s)
GM	General Motors
GNP	Gross National Product
GOSIP	Government Open System Interconnection Profile
HDTV	High Definition Television
HIPPI	High Performance Parallel Interface
IA	Implementation Agreement
IAB	Internet Activities Board

IBM	International Business Machines
IC	Integrated Circuit
ICC	Interstate Commerce Commission
IEC	International Electrotechnical Commission
IEEE	Institute of Electrical and Electronic Engineers
IFRB	International Frequency Reservation Board
IN	Intelligent Network
IO	Integrated Optical
IP	Intelligent Peripheral
ISDN	Integrated Services Digital Network
ISO	International Standards Organization
ISP	International Standardized Profiles
ITU	International Telecommunications Union
IVD	Integrated Voice and Data
IXC	Interexchange Carrier
kb/s	Kilobits per Second (10^3 b/s)
km	Kilometer
LAN	Local Area Network
LAP-B	Link Access Protocol B
LATA	Local Access Transport Area
LEC	Local Exchange Carrier
LEO	Low Earth Orbit
LFC	Local Function Capabilities
LLC	Link Level Control
M	Modem
MAC	Medium Access Control
MAN	Metropolitan Area Network
MAP	Manufacturing Automated Protocol
Mb/s	Megabits Per Second (10^6 b/s)
MCI	Microwave Communications Incorporated
MEO	Mid Earth Orbit
MF	Multiple Frequency
MGMT	Management
MHF	Medium High Frequency
MHS	Message Handling System
MILNET	Military Network
MIPS	Millions of Instructions per Second
MPC	Massively Parallel Computer
MUX	Multiplexer

NIST	National Institute of Standards and Technology
NNI	Network Node Interface
NREN	National Research and Education Network
NSF	National Science Foundation
NT	Network Termination
NTIA	National Telecommunications and Information Administration
OC	Optical Carrier
OIW	Open System Implementors Workshop
ONA	Open Network Architecture
OS	Operating System
OSI	Open Systems Interconnection
OSS	Operations Support System
PAN	Peculiar and Novel Service
PASS	Personal Access Satellite System
PBX	Private Branch Exchange
PC	Personal Computer
PCN	Personal Communications Network
PCS	Personal Communications System
PDN	Public Data Network
PDU	Protocol Data Unit
PLN	Private Line Network
PMD	Physical Medium Dependent
PMI	Physical Medium Independent
POSI	Pacific OSI (Japan)
POTS	Plain Old Telephone Service
PRI	Primary Rate Interface
PSTN	Public Switched Telephone Network
PTT	Postal Telegraph and Telephone
RBOC	Regional Bell Operating Company
RISC	Reduced Instruction Set Computer
ROM	Read Only Memory
RPOA	Recognized Private Operating Agency
SCP	Service Control Point
SCPC	Single Channel per Carrier
SDBN	Software Defined Broadband Network
SDH	Synchronous Digital Hierarchy
SDN	Software Defined Network

SDO	Standards Development Organization
SF	Single Frequency
SHF	Super High Frequency
SMDS	Switched Multimegabit Data Service
SMS	Service Management System
SNA	Systems Network Architecture
SONET	Synchronous Optical Network
SPAG	Standards Promotion and Applications Group
SPC	Stored Program Control
SS7	Signaling System No. 7
SSP	Service Switching Point
SST	Spread Spectrum Technology
STM	Synchronous Transfer Mode
STP	Shielded Twisted Pair
STS	Synchronous Transport System
SW	Switch
T-Carrier	Digital Transmission System
T1	T-Carrier Operated at 1.544 Mb/s
TASI	Time Assignment Speech Interpolation
Tb/s	Terabits per Second (10^{12} b/s)
TBD	To Be Determined
TCP/IP	Transport Control Protocol/Internet Protocol
TDM	Time Division Multiplexing
TDMA	Time Division Multiple Access
TE	Terminal Equipment
TIA	Telecommunications Industry Association
TMN	Telecommunications Management Network
TRW	Thompon Ramo Woolrich
TV	Television
TWP	Twisted Wire Pair
UPT	Universal Personal Telecommunications
UTP	Unshielded Twisted Pair
VAN	Value Added Network
VBR	Variable Bit Rate
VC	Virtual Channel
VCI	Virtual Channel Identifier
VLSI	Very Large Scale Integration
VP	Virtual Path
VPC	Virtual Path Connection

VPI	Virtual Path Identifier
VPLN	Virtual Private Line Network
VPN	Virtual Private Network
VSAT	Very Small Aperture Terminal
VT	Virtual Tributaries
WAN	Wide Area Network
WARC	World Administrative Radio Conference
WP	Wire Pair
WS	Work Station
XC	Exchange

Definitions

These definitions are taken from Federal Standard 1037B (1991), a glossary of telecommunication terms, wherever possible.

Asynchronous Transfer Mode (ATM) - A data-transfer mode in which a multiplexing technique for fast packet switching in CCITT broadband ISDN is used. This technique inserts information in small, fixed-size cells (32-120 octets) that are multiplexed and switched in a slotted operation, based upon header content, over a virtual circuit established immediately upon a request for service.

Asynchronous Transmission - Data transmission in which the instant that each character, or block of characters, starts is arbitrary; once started, the time of occurrence of each signal representing a bit within the character, or block, has the same relationship to significant instants of a fixed time frame.

Bandwidth-on-Demand - A method of transporting information according to each user's instantaneous need (e.g., see ATM and B-ISDN).

Boundary - An abstract separation between functional groupings of protocols. May or may not be a physical interface as well.

Broadband (wideband) - 1. An imprecise designation of a signal that occupies a broad frequency spectrum. Note: This term is often used to distinguish it from a narrowband signal, where both terms are subjectively defined relative to the implied context. 2. That property of any circuit having a bandwidth wider than normal for the type of circuit, frequency of operation, and type of modulation carried. Note: The term has many meanings depending upon application. In telecommunications, the term implies a service or system requiring transmission channels capable of supporting rates greater than 1.5 Mb/s.

Broadband ISDN (B-ISDN) - A CCITT proposed Integrated Services Digital Network offering broadband capabilities including many of the following features or services: (a) from 150 to 600 Mb/s interfaces, (b) using ATM to carry all services over a single, integrated, high-speed packet-switched net, (c) LAN interconnection, (d) the ability to connect LANs at different locations, (e) access to a remote, shared disc server, (f) voice/video/data teleconferencing from one's desk, (g) transport for programming services (e.g., cable TV), (h) single-user controlled access to remote video source, (i) voice/video telephone calls, and (j) access to shop-at-home and other information services.

Cell - 1. In cellular radio, the smallest geographic area defined for a certain mobile communication system. 2. In OSI, a fixed-length block labeled at the physical layer of the OSI reference model.

165

Cell-Relay - A multiplexed information transport method in which information is organized into fixed-length cells with an identifying header and transmitted according to users' instantaneous needs (e.g., see ATM).

Communications System - A collection of individual communication networks, transmission systems, relay stations, tributary stations, and terminal equipment capable of interconnection and interoperation to form an integral whole.

End System and End User - The ultimate source or destination for information transferred over a network.

Frame - In data transmission, the sequence of contiguous bits bracketed by and including beginning and ending flag sequence.

Implementation - Software and hardware that performs the logical functions defined by the network architecture.

Integrated Services Digital Network (ISDN) - An integrated digital network in which the same time-division switches and digital transmission paths are used to establish connections for different services. Note 1: Such services include telephone, data, electronic mail, and facsimile. Note 2: How a connection is accomplished is often specified. For example, switched connection, non-switched connection, exchange connection, ISDN connection. See also communications, electronic mail, integrated digital network.

Intelligent Network (IN) - A network that allows functionality to be distributed flexibly at a variety of nodes on and off the network and allows the architecture to be modified to control the services; [in North America] an advanced network concept that is envisioned to offer such things as (a) distributed call-processing capabilities across multiple network modules, (b) real-time authorization code verification, (c) one-number services, and (d) flexible private network services [including (1) reconfiguration by subscriber, (2) traffic analyses, (3) service restrictions, (4) routing control, and (5) data on call histories]. Levels of IN development are identified below:

--IN/1 A proposed intelligent network targeted toward services that allow increased customer control and that can be provided by centralized switching vehicles serving a large customer base.

--IN/1+ A proposed intelligent network targeted toward services that can be provided by centralized switching vehicles, e.g., access tandems, serving a large customer base.

--IN/2 A proposed, advanced intelligent-network concept that extends the distributed IN/1 architecture to accommodate the concept called "service independence." Note: Traditionally, service logic has been localized at individual switching systems. The IN/2 architecture provides flexibility in the placement of service logic, requiring the use of advanced techniques to manage the distribution of both network data and service logic across multiple IN/2 modules.

Interface - A concept involving the definition of the interconnection between two equipment items or systems. The definition includes the type, quantity, and function of the interconnecting circuits and the type, form, and content of signals to be interchanged·via those circuits.

Layered Architecture - Functional group of protocols that adheres to a logical structure of network operations.

Local Area Network (LAN) - A non-public data communication system, within a limited geographic area, designed to allow a number of independent devices to communicate with each other over a common transmission interconnection topology.

Metropolitan Area Network (MAN) - A loosely defined term generally understood to describe a network covering an area larger than a LAN. Note: It typically interconnects two or more LANs, operates at higher speed and may cross administrative boundaries.

Multimedia Communications - The field referring to the representation, storage, retrieval, and dissemination of machine-procurable information expressed in multimedia such as text, voice, graphics, images, audio, and video.

Network - 1. An interconnection of three or more communicating entities and (usually) one or more nodes. 2. A combination of passive or active electronic components that serves a given purpose.

Network Topology - The connecting structure, consisting of paths, switches, and concentrators that provides the communications interconnection among nodes of a network. Note: Two networks have the same topology if the connecting configuration is the same, although the networks differ in physical interconnections, distance between nodes, transmission rates, and signal types.

Open System - A system whose characteristics comply with specified standards and that therefore can be connected to other systems that comply with these same standards.

Open System Interconnection (OSI) - A logical structure for network operations standardized within the ISO; a seven-layer network architecture being used for the definition of network protocol standards to enable any OSI-compliant computer or device to communicate with any other OSI-compliant computer or device for a meaningful exchange of information.

Open System Interconnection (OSI) Architecture - Network architecture that adheres to that particular set of ISO standards that relates to Open Systems Architecture.

Overhead Bit - Any bit other than a user information bit.

Overhead Information - Digital information transferred across the functional interface separating a user and a telecommunication system (or between functional entities within a telecommunication system) for the purpose of directing or controlling the transfer of user information and/or the detection and correction of errors. Overhead information originated by the user is not considered as system overhead information. Overhead information generated within the system and not delivered to the user is considered as system overhead information.

Photonics - The field of telecommunications involving discrete packets of electromagnetic energy for switching and transmission.

Protocol - A set of unique rules specifying a sequence of actions necessary to perform a communications function.

T-Carrier - Generic designator for any of several digitally multiplexed telecommunications transmission systems.

Telecommunication - Any transmission, emission, or reception of signs, signals, writing, images, and sounds or intelligence of any nature by wire, radio, optical, or other electromagnetic systems.

Telecommunication Architecture - Within a telecommunication system, the overall plan governing the capabilities of functional elements and their interaction, including configuration, integration, standardization, life-cycle management, and definition of protocol specifications, among these elements.

Telecommunication Service - A specified set of user-information transfer capabilities provided to a group of users by a telecommunication system. The telecommunication service user is responsible for the information content of the message. The telecommunication service provider has the responsibility for the acceptance, transmission, and delivery of the message.

Synchronous Digital Hierarchy (SDH) - A newly adopted standard for multiplexing and interfacing signals for transmission over optical networks. Evolved from Synchronous Optical Network (SONET) developed in the United States.

Synchronous Transfer Mode (STM) - A proposed transport level, a time-division multiplex-and-switching technique to be used across the user's network interface for ISDN.

System - Any organized assembly of resources and procedures united and regulated by interaction or interdependence to accomplish a set of specific functions.

User - A person, organization, or other entity (including a computer or computer system), that employs the services provided by a telecommunication system, or by an information processing system, for transfer of information to others. Note: A user functions as a source or final destination of user information, or both.

User Information - Information transferred across the functional interface between a source user and a telecommunication system for the purpose of ultimate delivery to a destination user. Note: In data telecommunication systems, "user information" includes user overhead information.

Wide Area Network (WAN) - A physical or logical network that provides capabilities for a number of independent devices to communicate with each other over a common transmission-interconnected topology in geographic areas larger than those served by local area networks.

Printed and bound by CPI Group (UK) Ltd, Croydon, CR0 4YY

03/10/2024

01040335-0012